Lecture Notes in Biomathematics

Managing Editor: S. Levin

26

Myron Bernard Katz

Questions of Uniqueness
and Resolution in Reconstruction
from Projections

Springer-Verlag
Berlin Heidelberg New York

Lecture Notes in Biomathematics

Vol. 1: P. Waltman, Deterministic Threshold Models in the Theory of Epidemics. V, 101 pages. 1974.

Vol. 2: Mathematical Problems in Biology, Victoria Conference 1973. Edited by P. van den Driessche. VI, 280 pages. 1974.

Vol. 3: D. Ludwig, Stochastic Population Theories. VI, 108 pages. 1974.

Vol. 4: Physics and Mathematics of the Nervous System. Edited by M. Conrad, W. Güttinger, and M. Dal Cin. XI, 584 pages. 1974.

Vol. 5: Mathematical Analysis of Decision Problems in Ecology. Proceedings 1973. Edited by A. Charnes and W. R. Lynn. VIII, 421 pages. 1975.

Vol. 6: H. T. Banks, Modeling and Control in the Biomedical Sciences. V, 114 pages. 1975.

Vol. 7: M. C. Mackey, Ion Transport through Biological Membranes, An Integrated Theoretical Approach. IX, 240 pages. 1975.

Vol. 8: C. DeLisi, Antigen Antibody Interactions. IV, 142 pages. 1976.

Vol. 9: N. Dubin, A Stochastic Model for Immunological Feedback in Carcinogenesis: Analysis and Approximations. XIII, 163 pages. 1976.

Vol. 10: J. J. Tyson, The Belousov-Zhabotinskii Reaktion. IX, 128 pages. 1976.

Vol. 11: Mathematical Models in Medicine. Workshop 1976. Edited by J. Berger, W. Bühler, R. Repges, and P. Tautu. XII, 281 pages. 1976.

Vol. 12: A. V. Holden, Models of the Stochastic Activity of Neurones. VII, 368 pages. 1976.

Vol. 13: Mathematical Models in Biological Discovery. Edited by D. L. Solomon and C. Walter. VI, 240 pages. 1977.

Vol. 14: L. M. Ricciardi, Diffusion Processes and Related Topics in Biology. VI, 200 pages. 1977.

Vol. 15: Th. Nagylaki, Selection in One- and Two-Locus Systems. VIII, 208 pages. 1977.

Vol. 16: G. Sampath, S. K. Srinivasan, Stochastic Models for Spike Trains of Single Neurons. VIII, 188 pages. 1977.

Vol. 17: T. Maruyama, Stochastic Problems in Population Genetics. VIII, 245 pages. 1977.

Vol. 18: Mathematics and the Life Sciences. Proceedings 1975. Edited by D. E. Matthews. VII, 385 pages. 1977.

Vol. 19: Measuring Selection in Natural Populations. Edited by F. B. Christiansen and T. M. Fenchel. XXXI, 564 pages. 1977.

Vol. 20: J. M. Cushing, Integrodifferential Equations and Delay Models in Population Dynamics. VI, 196 pages. 1977.

Vol. 21: Theoretical Approaches to Complex Systems. Proceedings 1977. Edited by R. Heim and G. Palm. VI, 244 pages. 1978.

Vol. 22: F. M. Scudo and J. R. Ziegler, The Golden Age of Theoretical Ecology: 1923–1940. XII, 490 pages. 1978.

Vol. 23: Geometrical Probability and Biological Structures: Buffon's 200th Anniversary. Proceedings 1977. Edited by R. E. Miles and J. Serra. XII, 338 pages. 1978.

Vol. 24: F. L. Bookstein, The Measurement of Biological Shape and Shape Change. VIII, 191 pages. 1978.

Lecture Notes in Biomathematics

Managing Editor: S. Levin

26

Myron Bernard Katz

Questions of Uniqueness and Resolution in Reconstruction from Projections

Springer-Verlag
Berlin Heidelberg New York 1978

Editorial Board
W. Bossert · H. J. Bremermann · J. D. Cowan · W. Hirsch
S. Karlin · J. B. Keller · M. Kimura · S. Levin (Managing Editor)
R. C. Lewontin · R. May · G. F. Oster · A. S. Perelson · L. A. Segel

Author
Myron Bernard Katz
Department of Mathematics
University of New Orleans
New Orleans, Louisiana 70122/USA

AMS Subject Classifications (1970): 44A15, 92A05, 15A09, 15-01

ISBN 3-540-09087-8 Springer-Verlag Berlin Heidelberg New York
ISBN 0-387-09087-8 Springer-Verlag New York Heidelberg Berlin

This work is subject to copyright. All rights are reserved, whether the whole or part of the material is concerned, specifically those of translation, reprinting, re-use of illustrations, broadcasting, reproduction by photocopying machine or similar means, and storage in data banks. Under § 54 of the German Copyright Law where copies are made for other than private use, a fee is payable to the publisher, the amount of the fee to be determined by agreement with the publisher.
© by Springer-Verlag Berlin Heidelberg 1978
Printed in Germany

Printing and binding: Beltz Offsetdruck, Hemsbach/Bergstr.
2141/3140-543210

In memory of my father, Julius.
This is one of his accomplishments.

PREFACE

Reconstruction from projections has revolutionized radiology and has now become one of the most important tools of medical diagnosis - the E. M. I. Scanner is one example. In this text, some fundamental theoretical and practical questions are resolved.

Despite recent research activity in the area, the crucial subject of the uniqueness of the reconstruction and the effect of noise in the data posed some unsettled fundamental questions. In particular, Kennan Smith proved that if we describe an object by a C_o^∞ function, i.e., infinitely differentiable with compact support, then there are other objects with the same shape, i.e., support, which can differ almost arbitrarily and still have the same projections in finitely many directions. On the other hand, he proved that objects in finite dimensional function spaces are uniquely determined by a single projection for almost all angles, i.e., except on a set of measure zero. Along these lines, Herman and Rowland in [41] showed that reconstructions obtained from the commonly used algorithms can grossly misrepresent the object and that the algorithm which produced the best reconstruction when using noiseless data gave unsatisfactory results with noisy data. Equally important are reports in Science, [67, 68] and personal communications by radiologists indicating that in medical practice failure rates of reconstruction vary from four to twenty percent.

Within this work, the mathematical dilemma posed by Kennan Smith's result is discussed and clarified. A larger class of function spaces is shown to have a certain pathological property which maintains that any finite number of projections is not sufficient to specify the desired object function. As this property is not shared by finite dimensional function spaces, one is led to the study of discretized models even in the most abstract of approaches.

The main thrust of this work is concerned with the limitations on the application of reconstruction from projections imposed by practical diagnostic medical techniques. The particular finite dimensional function space examined is exactly the one most often used in brain scanning devices. Attention is paid to the empirical tradeoff between resolution and accuracy in projection data. Paramount, however, is the goal of optimal resolution in the reconstruction. This has clear diagnostic and theoretical significance since many medical anomalies require the finest image detail for their unambiguous identification. High resolution is also important to the justification of the finite dimensional reconstruction as a good approximation to the actual object. The medical context of this report is clarified in an appendix.

Two independent results indicate that a certain set of projection angles should be used. Theorem 1, in Chapter V, shows that this set of angles is the least demanding on the resolution of the projection data - thereby establishing that projection data with the highest statistical accuracy can be utilized at these angles. Theorem 2, in the sixth chapter, provides a simple formula specifying the resolution in the reconstruction which cannot be exceeded if (uniquely) determined information is required. That is, the extent to which an object is determined by its projection data depends on how one defines the approximation, i.e., the resolution of the reconstruction. The applicability of Theorem 2 requires that the projection angles come from the previously described set.

In Chapter VIII, it is shown that once physically justifiable assumptions are made, it is possible to predict how well a reconstruction approximates the original object function. An estimate which indicates the accuracy of the reconstruction is given.

The style of this work was chosen so that researchers in the general field of image reconstruction as well as interested physicians would find the ideas presented accessible. Although the mathematical treatment is not self-contained, nothing beyond second or third year college mathematics is assumed. In particular, the ideas developed in a course in linear algebra (or vector spaces) and the concept of a norm on a vector space are considered common or easily accessible knowledge. To facilitate reading, proofs of propositions and theorems are deferred until the ends of the chapters in which they occur.

This manuscript owes much to the help of many others. The original suggestion came from Hans J. Bremermann who followed up that suggestion with valuable insight and encouragement. Frequent consultation with Ole Hald, Paul Chernoff, William Bade, Kennan T. Smith, Alberto Grunbaum and Keith Miller provided support and infectious enthusiasm. Special efforts were made by Jaime Milstein, Kim Chambers and Nora Lee. I extend many thanks to David Krumme for his conscientious and indefatigable work in editing and criticizing every draft, and to Cherie Parker, for sustained moral support during a difficult period.

Myron Katz

Myron Katz

June, 1977

TABLE OF CONTENTS

CHAPTER		Page
I	DESCRIPTION OF THE GENERAL PHYSICAL PROBLEM	1
	The EMI Scanner - An Example of the Present State of the Art	8
	Reconstruction from Projections Models Many Physical Problems and Presents a Variety of Theoretical Questions	13
	The Difficulties Associated with the Theory of Reconstruction from Projections	17
II	BASIC INDETERMINACY OF RECONSTRUCTION	21
	Theoretical Background	21
	First Theoretical Result with Practical Significance	23
	The Significance of the Nullspace	25
	Does There Exist a Restriction on the Domain of $P_{\{\theta\}}$ Which Makes $N = \{0\}$?	27
	Conclusions to Chapter II	30
	Proofs of Results Stated in Chapter II	31
III	A RECONSTRUCTION SPACE WHICH DOES NOT CONTAIN THE OBJECTIVE FUNCTION	38
	A Reconstruction Space Based on the Fourier Transform	38
	Description of Our Choice of the Reconstruction Space	41
	Resolution of a Reconstruction $\equiv$ Picture Resolution	43
IV	A MATRIX REPRESENTATION OF THE PROBLEM	44
	Proofs of Propositions Stated in Chapter IV	52
V	RESOLUTION IN THE PROJECTION DATA	53
	Projection Angles Affect the Required Resolution	53
	Farey Series and Projection Angles	54
	Significance of the Farey Projection Angles	56
	Proofs of Results Stated in Chapter V	57

CHAPTER		Page
VI	RESULTS ESTABLISHING THE UNIQUENESS OF A RECONSTRUCTION	62
	Interpretation of the Two Uniqueness Results: Proposition VI.2 and Theorem 2	65
	There Is in Practice a Limitation on the Resolution in $P_\theta f$	67
	Explanation of Theorem 2	70
	Uniquely Determined Picture Resolution	75
	Proofs of Results Stated in Chapter VI	80
VII	DEALING EFFECTIVELY WITH NOISY DATA	91
	Physical Justification of Importance and Sources of Noise	92
	The Effect of Noisy Data on the Uniqueness of a Reconstruction	92
	The Effect of Noise on the Consistency of the Data	96
	The Use of Least Squares - Advantages and Difficulties	97
	Statistical Considerations Relevant to the Use of Least Squares	100
	Optimizing the Stability of the Estimate of the Unknown Reconstruction	102
	Choosing the Best Projection Angles	103
	Conclusions to Chapter VII	108
APPENDIX TO CHAPTER VII - STATISTICAL REFERENCE MATERIAL		109
VIII	HOW A RECONSTRUCTION APPROXIMATES A REAL LIFE OBJECT	113
	Assumptions with Their Justifications	114
	Consequences of the Assumptions	118
	Estimating $\|h - f\|_{L^2}$, i.e., How close is the obtained reconstruction to the unknown objective function?	124
	Significance and Applications of the Estimate of $\|h - f\|_{L^2}$	127
	Conclusions	128
	Proofs of Propositions stated in Chapter VIII	131

CHAPTER		Page
IX	A SPECIAL CASE: IMPROVING THE EMI HEAD SCANNER	138
	The Use of Purposefully Displaced Reconstructions	140
	Theorem 2 Applied to Four Sets of Purposefully Displaced Projection Data	140
	Estimating the Accuracy of a 74 x 74 Reconstruction, h_{74}	142
	Obtaining a Uniquely Determined Reconstruction with 1 mm Resolution from 1 mm Resolution Projection Data	143
	Conclusions	144
X	A GENERAL THEORY OF RECONSTRUCTION FROM PROJECTIONS AND OTHER MATHEMATICAL CONSIDERATIONS RELATED TO THIS PROBLEM	145
	A General Theory of Reconstruction from Projections	145
	Other Mathematical Considerations Related to Reconstruction from Projections	149
APPENDIX - MEDICAL CONTEXT OF RECONSTRUCTION FROM PROJECTIONS		154
	The Interaction of X-rays with Matter	155
	The Meaning of a Projection	157
	Thickness of the Slice	157
	Types of Detectors	158
	Parallel and Fan-beam Techniques	158
	Resolution of the Data	159
	X-Ray Exposure	160
	Miscellaneous Aspects of Data Collection	160
	The EMI Example	161
	Algorithms	162
	Representation of a Reconstruction	168
	The Diagnosis Problem	169
REFERENCES		170

CHAPTER I

DESCRIPTION OF THE GENERAL PHYSICAL PROBLEM

Consider an object with varying density (of mass). For example: A human body has bones, lungs, muscles and etc. It is possible to make the "inner structure" visible through the use of x-rays, but we only get a projection. How could the actual three dimensional mass distribution be determined?

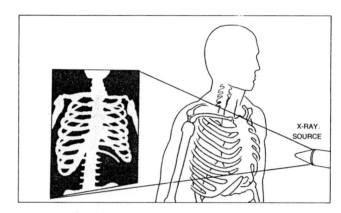

Figure I.1
[34]

CONVENTIONAL X-RAY PICTURE is made by allowing the X rays to diverge from a source, pass through the body of the subject and then fall on a sheet of photographic film.

This **report** attacks the general problem of reconstruction from projections. That is, the problem of determining the inner structure of an object from a finite number of projections of that object. In addition, the **exposition is** concerned with the practical problem of reconstruction from projections, which is subject to inaccurate data collection and a limited amount of information in each projection. The problem of determining the mass distribution within a human body with x-ray projection data is important to medical diagnosis. It is used as a typical example of a more general class of reconstruction-from-projections problems.

From the outset this **report** considers all problems to be two-dimensional. A three-dimensional problem can be reduced to a two-dimensional problem by recognizing that if a horizontal (planar) cross-section of the mass distribution is known at every height then all of the three-dimensional information is readily available. We are now considering projection data to be the information within the radiograph along a line which is contained in the cross-sectional plane of interest. By this change in the definition of the projection data, we can convert a three-dimensional problem into a two-dimensional problem.

When x-rays pass through matter, they are **attenuated roughly** in proportion to the density of that matter. Thus if we record the intensity of x-rays that have passed through an object, we get a projection of the mass density. If the purpose of such an experiment is to determine a cross-section of the density distribution in a man's head, then the physical problem is reduced to recovering that distribution from a finite number of radiographs.

In the beginning, it is well to translate this problem into a clear mathematical question. We describe the given mass distribution by its density at each point. Let f be a real-valued function defined on a plane. (f actually represents the coefficient of linear x-ray attenuation at each point within the particular cross-sectional plane.)

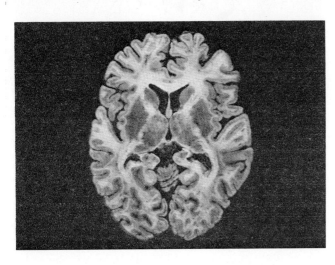

Figure I.2
[34]
A photograph of a cross-section of a dissected normal brain.

Because f represents a physical object, it is a non-negative function which is considered to be zero in air (since attenuation of x-rays by air is small). f will be called the *objective function,* inasmuch as recovering f is the goal of the analysis. In particular, this means that f is only non-zero on a bounded subset of the plane, i.e., f has compact support. By rescaling the plane, if necessary, another simplifying assumption can be made: f is non-zero only within the unit square I^2, i.e., $f(x,y) \neq 0$ implies that $0 \leq x,y \leq 1$.

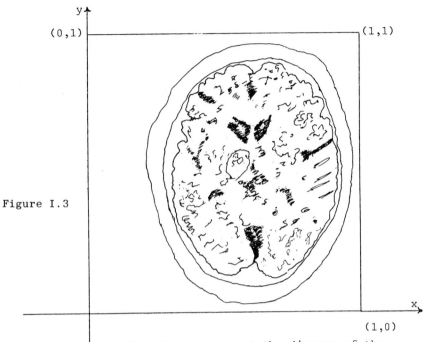

Figure I.3

f is only non-zero at the diagram of the cross-section of the head.

Once we have chosen a coordinate system a *projection* is characterized by an *angle,* θ , which we measure in a counter-clockwise direction from the x-axis.

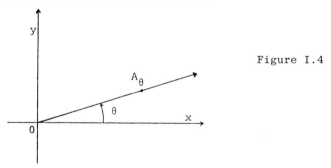

Figure I.4

Let A_θ be any point on the ray determined by θ . The ray, $\overrightarrow{OA_\theta}$, is completely determined by θ and in turn specifies $\theta^\perp$, the line perpendicular to $\overrightarrow{OA_\theta}$ and passing through 0.

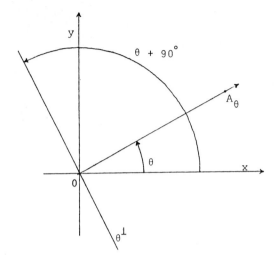

Figure I.5

A *projection of* f *in the direction* θ is given by:

<u>I.1</u> $$P_\theta f(v) = \int_{-\infty}^{+\infty} f(v + tA_\theta)dt, \text{ for } \mathbf{v} \text{ in } \theta^\perp.$$

That is, $P_\theta f$ is a new function which is defined on the line $\theta^\perp$ and has the property that its value at any point v in $\theta^\perp$ only depends on that part of f defined on the line $\{v + tA_\theta : t \text{ is in } R\}$.

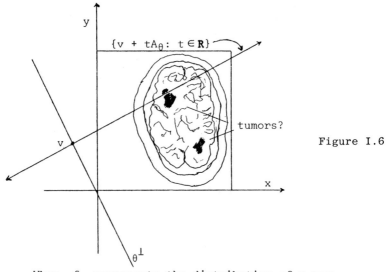

Figure I.6

When f represents the distribution of x-ray attenuation in a cross-section of a human head, then $P_\theta f$ is usually just an x-ray or radiograph of that head read along the appropriate line on the photographic film. (In actual practice the optical density measured records $\exp(P_\theta f(v))$ so that the original data must be converted with the logarithm to get the desired projection information. See pages 155-157.)

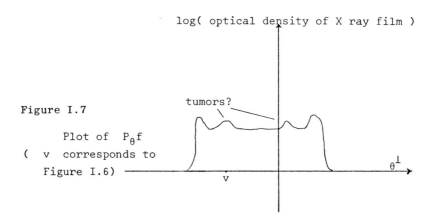

Figure I.7

Plot of $P_\theta f$
(v corresponds to Figure I.6)

For notational convenience, it can be assumed that A_θ has length one, so that A_θ is uniquely determined by θ. It is reasonable to drop the name A_θ in favor of $\vec{\theta}$. This should cause no confusion since the meanings of θ and $\vec{\theta}$ are naturally related:

(1) θ is an *angle* measured in the counter-clockwise direction from the x-axis, and (2) $\vec{\theta}$ is the *vector* with length one which makes the same angle with the x-axis.

With this convention, the fundamental definition of a *projection of* f *in the direction* $\vec{\theta}$ becomes:

I.2 $$P_\theta f(v) = \int_{-\infty}^{+\infty} f(v + t\vec{\theta}) \, dt, \quad \text{for } v \text{ in } \theta^\perp.$$

In the present introductory discussion a large class of mathematical questions like the well-definedness of P_θ is postponed, but the rigorous reader should be satisfied with the single condition that f be piecewise continuous.

Because of time and **numerous** other **considerations,** any physical measurement only contains a finite amount of information. The time limitation translates into very significant economic **constraints:** the present imaging techniques cost about $200 to the patient. Furthermore, some organs, like the breast, are very sensitive to radiation damage. Because of the need to protect the patient from an excessive dose and because of monetary factors, it is desirable to take as *few* projections as possible as is consistent with the *goal* to get a medically acceptable reconstruction.

For notational convenience, the letter m will be used to indicate the number of projection angles used, i.e., the observed data is $\{P_{\theta_1}f, P_{\theta_2}f, \ldots, P_{\theta_m}f\}$, corresponding to the projection angles $\{\theta_1, \theta_2, \ldots, \theta_m\}$. The general mathematical question is to determine f from $\{P_{\theta_1}f, P_{\theta_2}f, \ldots, P_{\theta_m}f\}$.

The EMI Scanner - An Example of the Present State of the Art

Determining the coefficient of x-ray absorption everywhere within a cross-section of a human head from a finite number of radiographs is an example of the type of problem addressed in this report. This problem is particularly difficult because variations of density are small. It is an important problem since the solution would make it possible to detect brain tumors. There are many other physical problems which can be modeled with the same mathematical machinery. Examples of these problems range from radio astronomy to plasma physics, and from study of air flow in a wind tunnel to electron microscopy. (See later discussion about determining the structure of protein molecules with projection data obtained from an electron microscope, p.14.) Medical practitioners have become interested in reconstruction from projections surpassing its long standing use in nuclear medicine: a new device called the EMI Scanner has proven to be generally reliable in producing mass density distributions of cross-sections of human brains from x-ray projection data. So that at least one example is developed in full, the EMI Scanner is explained.

The EMI Scanner processes x-ray projection data and produces a reconstruction (i.e., finite dimensional approximation) of the actual mass distribution of a cross-section of the patient's head. The first problem to be overcome is the method of data collection. This can be accomplished by scanning the patient and feeding this data into a computer as in Figure I.8.

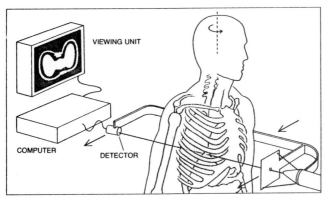

Figure I.8
[34]

RECONSTRUCTION FROM PROJECTIONS is made by mounting the X-ray source and an X-ray detector on a yoke and moving them past the body. The yoke is also rotated through a series of angles around the body. Data recorded by detector are processed by a special computer algorithm, or program. Computer generates picture on a cathode-ray screen.

In the particular case of the EMI, the scanning is restricted only to the head. The next photograph shows a patient lying on a table so that her head can be sampled with the scanning equipment of the EMI.

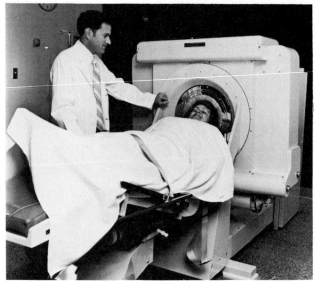

Figure I.9
[1] [34]

SCANNER SAMPLES RAY SUMS AT 160 POINTS along each projection; in five and a half minutes 180 projections are taken at one-degree intervals around the patient's head.

In this case, the head is held in place by a water bag. Then a scanning detector system takes a projection at each of 180 different angles in 1° increments. Each projection is actually 160 different measurements so that 160 × 180 observations are inputed into a computer for processing.

The x-ray equipment will have inputed 11,200 (160×180) different numbers which represents $\{P_{\theta_1}f, P_{\theta_2}f, \ldots, P_{\theta_{180}}f\}$ where $\theta_1 = 1°$, $\theta_2 = 2°$, ..., $\theta_{90} = 90°$, ..., $\theta_{180} = 180°$. In this case, however, each projection has been divided into 160 different data collection regions.

The next step in the process is the use of a computer implemented algorithm which transforms the projection data into the desired reconstruction. For some time, the algorithm called the Algebraic Reconstruction Technique or *ART* developed by Herman, Gordon, and Bender [32], was used to perform the desired transformation. Later the *convolution* method of Ramachandran and Lakshminarayanan [63], was implemented. [41] (See pp 162-168 for more explanation.)

Finally, the reconstruction is displayed on a cathode ray tube or T.V. screen. Figure I.10 compares a reconstruction with an actual cross-section; notice that the obtained reconstruction is far less complex than the part of the brain that it is supposed to represent.

Figure I.10 [34]

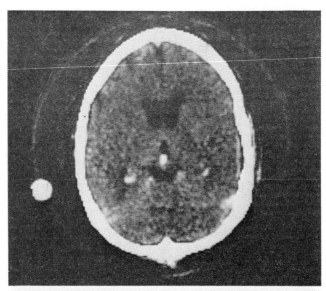

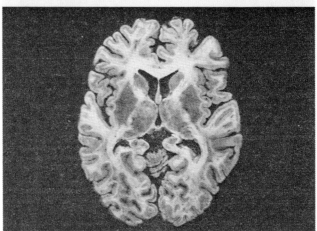

COMPARISON OF TWO CROSS SECTIONS, one an image of a normal human brain obtained by the EMI scanner (*top*) and the other a photograph of a section of a dissected normal human brain (*bottom*), illustrates the operation of the scanner. The anatomical details of the two brains can be readily compared. The reconstructed spot that appears to the left of the scanner image corresponds to a plastic rod that is used for calibrating the X-ray densities in the image. The scanner image of the head consists of 11,200 (160 × 160) picture elements. The image was made in the course of a study by D. F. Reese of the Mayo Clinic.

Reconstruction from Projections Models Many Physical Problems and Presents a Variety of Theoretical Questions

The problem of reconstruction from projections occurs in many fields. Usually projection data is obtained by particles or waves passing through the object to be investigated. In radio astronomy, radio waves are being emitted from distant astronomical regions; the resulting data can be plotted to form a projection. In plasma physics, the electron number density can be sampled by observing the variation in optical refractive index when visible light is passed through the plasma field. Gamma ray emitting isotopes are injected into patients for the purposes of nuclear medicine; gamma detectors measure projections that are attenuated by the density of the intervening body tissue. Heavy ions are utilized instead of x-rays for obtaining projection data on the same problem for which the EMI was designed. Positrons have been used to generate simultaneous and oppositely directed gamma rays within patients; this technique is yet another rival for the EMI. Sound waves transmitted through tissue produce projection data of living patients. The last four examples are all in competition with the EMI for solving important medical diagnostic problems. Finally, an important technique in biological research concerning the electron microscopy of protein molecules will be discussed in the third paragraph following. [46]

The important feature common to reconstruction from projection problems is that a measurement made externally actually represents (or can be interpreted to represent) the sum of all individual measurements along a line through the object.

As a further example consider the mechanism of a (transmission) electron microscope. In that case the object (sample) is frequently sliced very thinly and then bombarded with high energy electrons. The electrons that penetrate through the sample have done so (usually) because they did not interact with matter. Therefore, to a good approximation, the configuration of electrons passing through the sample represents a projection of all the density encountered along each pathway of bombarding electrons, i.e., along lines.

The general field of reconstruction from projections has particularly benefited by the work of scientists who probed into the fine structure of molecules with the electron microscope. In their 1970 article on this subject, Crowther, DeRosier, and Klug [20] demonstrated an early technique for generating reconstructions; they even had some interesting resolution results which will be discussed in Chapter VI (p. 77). More recently, R. Henderson and P.N.T. Unwin were able to obtain 7-A resolution in a reconstruction of a protein in the cell membrane of *Halobacterium halobium*. [38] Their success depended on the clever utilization of the projection theorem which is presented among the proofs in chapter II (p.31) [71] Their method of data collection was first suggested by R. Gleaser. [30]

Richard Gordon of the Image Processing Unit, National Cancer Institute, National Institute of Health mentioned in his preface to a conference on this subject in August, 1975 [46] a wide gambit of fields where reconstruction from projections has become a valuable tool. He also showed that the many mathematical and statistical papers on the subject demonstrated that theoretical questions persist.

The transformation of projection data into a reconstruction has been attacked by many algorithms. An early technique was suggested by two radio astronomers, R. N. Bracewell and A. C. Riddle [8]. Their method, later rediscovered by G. N. Ramachandran and A. V. Lakshminarayanan [63], has since been called the convolution technique. This technique is a one step method and is generally fast in running when used on a computer.

A commonly used algorithm is ART, for Algebraic Reconstruction Technique. It was first proposed by Gordon, Bender and Herman [32] and later improved by Herman, Lent and Rowland [40]. This technique or one of its descendents is frequently applied since it has been used for some time and requires a minimum of program length and computer storage space. This technique is iterative.

Another algorithm was introduced by Gilbert [29], it is called SIRT, for simultaneous iterative reconstruction technique; it is also iterative but effectively performs many iterations of ART in each of its iterations.

A technique called ILST, for iterative least squares technique, was first proposed by Goitein [31] and later developed by Budinger [13].

An extended list of such references is included within the bibliography.

Although there has been much qualified success in obtaining good reconstructions, only a few theoreticians have dealt with the fundamental questions. In 1917, J. Radon [28] showed that all the projection data in all the directions would uniquely determine a function in a general class of functions. Since then it was shown that if f is square-summable and has compact support then any infinite number of radiographs are sufficient to determine it. [71] But it wasn't until Kennan T. Smith, [71] showed that *if f is an arbitrary infinitely differentiable function with compact support, then it is completely undetermined by any finite number of its projections,* that theoretical results began to have great impact on the direction of current research. D. Solomon, [73], [37] examined theoretical questions along these lines.

Aside from the more general theoretical problems, researchers have been concerned with the uniqueness of a reconstruction. Oppenheim and Mersereau [55] have an interesting result which is later quoted in this treatise (p. 63). Papers by Shepp and Logan [50] and DeRosier and Klug [21] have attacked this question.

An extended list of such references is included within the bibliography.

The Difficulties Associated with the Theory of Reconstruction from Projections

Since it must be accepted that only a finite number of projections is practically available, the theoretical results which require more projection data are not applicable.

The first question one asks is: Can an arbitrarily finely structured objective function be determined by a finite number of projections? At first glance, the answer seems to be no, since Kennan Smith's result says that non-uniqueness persists for infinitely differentiable functions with compact support. On closer examination of his result, it is not clear whether an added assumption (such as f is never greater than some reasonable bound) could dramatically change the consequences. This issue is discussed in the next chapter.

Because there is no practical need for finding all of the fine structure of an arbitrary objective function, scientists have been quite happy to restrict their attention to finding a finite dimensional reconstruction of the objective function -- by reconstructing it like a television picture on a finite grid, by a Fourier reconstruction with finitely many terms or etc.

For notational convenience, let *h be a reconstruction of f* so obtained, i.e., h is a member of a finite dimensional reconstruction space. (This notation will be used consistently.) If h is to be considered useful, it must be sufficiently well defined so that it can be inferred to be close to f. Within this **report**, the finite dimensional reconstruction space used is the same one utilized by television or the human eye, namely a finite rectangular grid of small squares. The individual elements or grid squares are called *pixels*. Hence, we call this space a *pixel reconstruction space*, and we equate the *picture resolution* with the size of a pixel.

The next issue, therefore, is whether the projection data uniquely determines h. Oppenheim and Mersereau have shown that h can always be determined by a single projection if the data has sufficiently fine resolution. However, the *resolution in the projection data* required may be grossly in excess of what is available. Real data have limited resolution and are subject to noise. Limitations on the *resolution in the projection data* arise from the fact that projection values at two nearby points may not be distinguishable, i.e., due to physical effects in the detection of x-ray intensities along a projection line the intensities at nearby points cannot be distinguished. (For example, in the EMI scanner intensity is measured at points 1 mm apart, this is typical for the state of the art.) We give results which predict the optimal *picture resolution* which can be obtained with a particular quality of projection data. In particular, if for a given resolution in the projection data, picture resolution attempted is too fine, then the *uniqueness of a reconstruction* is lost. Finally, we can completely characterize the family of all equally accurate reconstructions (i.e., reconstrucions which have the same projection data) which can be generated if the picture resolution attempted is not compatible with that allowed by these results.

Noise in the data can dramatically effect the choice of a reconstruction. It is not enough that h is uniquely determined by noiseless projection data, since that property does not predict how much h can change as an effect of a small change in the projection data. The question is: once you have done what is physically possible to minimize noise, what is its effect upon your reconstruction? How can we minimize its effect? What is the best picture resolution which can be obtained in the presence of noise? How should the projection angles be chosen? These problems are treated statistically in Chapter VII, and an algorithm is presented which allows for the choice of the optimal projection angles so that stability in the reconstruction is maximized.

So far we have assumed that the objective function belongs to a pixel reconstruction space. Suppose, however, that the objective function, f, does not satisfy this simplifying assumption, and suppose that we take the projection data of f and obtain a reconstruction h as before [on a grid], how close is h to f? This question involves a choice of a measure of closeness, i.e., the topology of a function space that contains both f and h. We give estimates in the L^2 norm (Chapter VIII). In spite of the fact that this is a very fundamental question, this has not (to our knowledge) been previously discussed.

The arguments presented may be important to the description of the problem of reconstruction of cross-sections of human heads to find brain tumors. The mathematical problems addressed in this thesis were chosen by the desire to solve problems that have already been shown to be important for physical reasons. Mathematical work on interesting problems which have been stimulated by this project will be presented in later papers.

CHAPTER II

BASIC INDETERMINACY OF RECONSTRUCTION

The history of determining an arbitrary function from its projections is at least a half century old. During that time, mathematicians have been able to provide some encouraging results when they assumed (the unrealistic situation) than an infinite amount of projection data is available. However, it was not until 1973 that fundamental work addressed the more practical question which follows from the fact that only a finite number of projections can ever become available. This recent result, of Kennan T. Smith, demonstrates that even with very strong restrictions on the objective function, *a finite number of radiographs seem to say nothing at all about determining this function.* This chapter shows that the basic indeterminacy demonstrated by Kennan Smith does not depend on his choice of restrictions, i.e., the non-uniqueness is independent of almost any choice of the space from which the objective function comes (as long as it is infinite dimensional). This chapter concludes that *some change in the mathematical model is needed.*

Theoretical Background

In 1917, an Austrian mathematician, J. Radon, published a paper that demonstrated that an objective function can be recovered from all of its projections. He showed that if f

is infinitely differentiable and rapidly decreasing,* then f
is determined by $\{P_\theta f: 0 \leq \theta \leq 180°\}$. [28] This is
certainly an encouraging result since **scientists would like**
to know that the information for which they are searching is
determined. Although it is true that an objective function
can be assumed to be infinitely differentiable and rapidly
decreasing, there certainly is no way to get projection data
on lines where the lines form a continuum of distinct lines.

Some progress was made with the following proposition,
but it still suffers from the same difficulty of requiring
too much projection data.

Proposition 0: If f is in $L_0^2(R^2)$ then f is determined
by an infinite set of projections.** [71]

This proposition improves that of J. Radon by further
restricting the objective function to compact support. It
is also encouraging since it requires even less projection
data than Radon's result, but it is still, nevertheless,
impractical in its assumptions.

---

*f *is infinitely differentiable and rapidly decreasing* means
that for every k > 0, and every positive integer p we have:
$\lim_{|z|\to\infty} |z|^k f^{(p)}(z) = 0$, where $f^{(p)}$ is the pth derivative
of f.

**f *is in* $L_0^2(R^2)$ means that f is non-zero on a bounded
subset of R^2 and $\int_{R^2} |f|^2 dz$ is finite.

First Theoretical Result with Practical Significance

An important, practical breakthrough in this field occured in 1973, when Kennan T. Smith proved the following theorem. His result has physical consequences because of the realistic nature of its hypotheses; he assumes that no more than a finite number of projections is known.

Theorem   (K. T. Smith)   A finite set of projections tells nothing at all.   (Which is his paraphrase for:)

Theorem   (K. T. Smith)   Suppose we have an infinitely differentiable object f and a finite number of directions. Then there is a new infinitely differentiable object f' with exactly the same shape (i.e., the supports of f and f' are the same), exactly the same projections from these directions, and completely arbitrary on any compact subset of the interior of the support f. [71]

Smith's interpretation of his result is that a reconstruction cannot be proven reliable unless some extra *a priori* information about the objective function is known.[*] In particular, his result states that it is not sufficient to know that f is infinitely differentiable with compact support.

---

[*] Personnal communication.

At first glance, his result suggests that reconstruction of an object from its projections is practically impossible, but how could that be true while the EMI is enjoying a good deal of reliability? The difference between the pessimistic outlook suggested by Kennan Smith's theorem and the obvious practical success of the EMI must depend, therefore, on some added information about a cross-section of a human head which was not assumed by Kennan Smith. It is certainly true that more can be said about a cross-section than the fact that it can be represented by an infinitely differentiable function with compact support. Directed by these considerations, Chapters VIII and X will answer the following questions: What is the nature of the *a priori* information which changes the mathematical question from hopeless indeterminacy to a practically resolvable problem? More significantly, how can this *a priori* information be utilized?

So far, mathematics has shown that merely knowing that the objective function is infinitely differentiable with compact support and has projection data in a finite number of directions is insufficient for uniquely determining f. To reconcile the EMI's success, intuition suggests that if f is also assumed to be bounded; a uniqueness theorem might result. Some result might follow if it could be assumed that the first derivative of f were bounded; similarly, a restriction on higher derivatives of f could be fruitful. Instead of following this pathway, the next section presents some fundamental results which further clarify the nature of the indeterminacy revealed by Kennan Smith's theorem.

The Significance of the Nullspace

The heart of the matter lies in the method employed by Kennan Smith to prove this theorem; he used the nullspace of the set of projection transformations, $\{P_{\theta_1}, P_{\theta_2}, \ldots, P_{\theta_m}\}$. The g of his proof has the property that $P_{\theta_j}(g) = 0$ for all j= 1, 2, ..., m. It is immediately clear that if **some** such non-trivial g exists, then f **can not be uniquely** determined by its projection data.

In the sequel the following extension of the definition of P_θ is used.*

Definition: $P_\theta : C_0(R^2) \to C(\theta^1)$ by $P_\theta f(v) = \int_{-\infty}^{+\infty} f(v+t\vec{\theta})dt$, where $C_0(R^2)$ is the set of continuous functions defined on R^2 with compact support and $C(\theta^1)$ is the set of continuous functions defined on θ^1.

When we are talking about more than one projection angle use:

---
*When we are discussing reconstruction spaces, i.e., after Chapter II, the domain of P_θ must be extended to include elements of the reconstruction space.

Definition: Let $\{\vec{\theta}_j\}_{j=1}^{m} \subseteq S^1$ be given. Then $P_{\{\theta\}} \equiv P_{\{\theta_j\}} \equiv P_{\{\theta_j\}_{j=1}^m}$ is defined by

$$P_{\{\theta\}}: C_0(R^2) \to \bigoplus_{i=1}^{m} C(\theta_i^1), \quad \text{i.e.,}$$

$$P_\theta f(v_1, v_2, \ldots, v_m) \equiv (P_{\theta_1} f(v_1), P_{\theta_2} f(v_2), \ldots, P_{\theta_m} f(v_m)) \ . \overset{*}{}$$

Let $N \equiv$ nullspace of $P_{\{\theta\}} = \bigcap_{j=1}^{m}$ nullspace of P_{θ_j}. Then, $P_{\{\theta_j\}_{j=1}^m}$ has an inverse on $C_0(R^2)$ if and only if N is the zero subspace. (Notice that N depends on $\{\vec{\theta}_j\}_{j=1}^m$.) Therefore, the reconstruction problem depends on N. We shall call any element of N a *ghost*. In case more than one choice of $\{\vec{\theta}_j\}_{j=1}^m$ is possible, the terminology: *g is a ghost with respect to* $\{\vec{\theta}_j\}_{j=1}^m$ is used. The following proposition lists many properties of N.

Proposition II.1 Let g be a ghost with respect to $\{\vec{\theta}_j\}_{j=1}^k$; let $\vec{\theta}_{k+1} \in S^1$; and let p be any element of $C_0(R^2)$, then:

$\overset{*}{}$Let $U_1, U_2, \ldots, U_n$ be vector spaces over R, then $\bigoplus_{i=1}^{n} U_i = \{(u_1, u_2, \ldots, u_n): u_i \in U_i\}$ is called the *direct sum* of the vectors spaces $U_1, U_2, \ldots, U_n$.

$S^1 = \{z \in R^2: |z| = 1\}$.

a) $\sum_{i=1}^{n} a_i \, g(z + z_i)$ is a ghost with respect to $\{\vec{\theta}_j\}_{j=1}^{k}$, where $a_i \in R$ and $z_i \in R^2$;

b) $g(z) - g(z + t_0 \vec{\theta}_{k+1})$ is a ghost with respect to $\{\vec{\theta}_j\}_{j=1}^{k+1}$, for any $t_0 \in R$;

c) $p * g$ is a ghost with respect to $\{\vec{\theta}_j\}_{j=1}^{k}$, where $p * g$ means convolution; and

d) $g(sz)$ is a ghost with respect to $\{\vec{\theta}_j\}_{j=1}^{k}$, where $s \in R$.

The definition of P_{θ_j} as well as its linearity properties immediately imply a), b), and d), while c) is an easy consequence of the projection theorem. (This theorem is presented among the proofs which appear at the end of this chapter.)

Does There Exist a Restriction on the Domain of $P_{\{\theta\}}$ Which Makes N = {0} ?

The next proposition introduces what seems to be the most fundamental component of N: K is a recursively defined element of N which exploits the properties of N. Its definition answers the question: construct in the simplest way possible a ghost w.r.t. $\{\vec{\theta}_j\}_{j=1}^{m}$ from an arbitrary element of $C_0(R^2)$. More importantly, it addresses the question: does there exist a restriction on the domain of $P_{\{\theta\}}$ which makes $N = \{0\}$?

Proposition II.2 Let X be any vector space of real valued functions defined on R^2, and $\{\vec{\theta}_j\}_{j=1}^{m} \subseteq S^1$. If there exists a non-zero $K \in X$ with the recursive definition:

i) $h_0 \in X$ and h_0 is in the domain of P_{θ_j} for all j,

ii) $h_j(z) = h_{j-1}(z) - h_{j-1}(s + t_j \vec{\theta}_j)$ for some $t_j \in R$,

iii) $K = h_m$,

then $K \in N \neq \{0\}$.

Corollary: Any restriction of the domain of $P_{\{\theta\}}$ which has the property that a non-zero K is contained in the domain of $P_{\{\theta\}}$ for every finite subset of S^1, will not allow an improvement of Proposition 0, i.e., reconstruction with a finite number of projections is not unique.

To complete this description of N, a few examples of K are given. Let $\theta_1 = 0$, $\theta_2 = \pi/2$, $\theta_3 = \arctan(1)$, $\theta_4 = \arctan(-1)$, and $\theta_5 = \arctan(\frac{1}{2})$. Figures II.1 through II.5 exhibit K for $\{\vec{\theta}_j\}_{j=1}^{m}$ when $m = 1, 2, \ldots, 5$; $t_1 = t_2 = 1$, $t_3 = t_4 = \sqrt{2}$, and $t_5 = \sqrt{5}$; and h_0 is the characteristic function on the unit square. (The choice of the t_j's allow K to be represented on a rectangular grid.)

Figure II.1

```
+1-1
```

Figure II.2

```
-1+1
+1-1
```

Figure II.3

```
  +1-1
-1    +1
+1-1
```

Figure II.4

```
  +1-1
-1    +1
+1    -1
  -1+1
```

Figure II.5

```
          -1+1
    +1          -1
-1     -1+1     +1
+1          -1
  -1+1
```

It is very difficult to describe a function space which does not fall victim to the corollary because the properties of K are intrinsic to R^2 and integration on R^2. For example, the fact that $h_5(z)$, in the definition of K, is a ghost with respect to $\{\vec{\theta}_5\}$ is a consequence of the fact that integration is left translation invariant on the locally compact group R^2, i.e., for $v \in \theta_5^{\perp}$

$$
\begin{aligned}
P_{\theta_5}(h_5)(v) &= P_{\theta_5}(h_4(z) - h_4(z + t_5\vec{\theta}_5))(v) \\
&= \int_{-\infty}^{+\infty} (h_4(v + t\vec{\theta}_5) - h_4(v + t_5\vec{\theta}_5 + t\vec{\theta}_5))\, dt \\
&= \int_{-\infty}^{+\infty} h_4(v + t\vec{\theta}_5)dt - \int_{-\infty}^{+\infty} h_4(v + t_5\vec{\theta}_5 + t\vec{\theta}_5)\, dt \\
&= 0
\end{aligned}
$$

It seems that the only way to exclude K is to make it impossible for $h_{j-1}(z)$ and $h_{j-1}(z + t_j \vec{\theta}_j)$ to be permissible functions simultaneously when $t_j \neq 0$. The reader can see that many different sets of restrictions can be made on f, the objective function, without changing the effect of Corollary to Proposition II.2.

Conclusions of Chapter II

In conclusion, it seems that indeterminacy persists in a very large class of function spaces within which f could be defined. In restrospect, we can return to the implications of Kennan Smith's theorem. As was mentioned in that section, more *a priori* information must be given if uniqueness is to be established. Certainly the question of whether we have a *well-posed*[*] problem is equally important and must also depend on a greater amount of information about f. Therefore, at the end of this chapter the reader should understand that *some change in the mathematical model is needed* if the problem of reconstruction from projections is to become practically resolvable.

---

[*]Since our problem is linear, we can use the following definition. A problem is *well-posed* if for any given data there is a unique solution determined by the data and the solution depends continuously on the data.

Proofs of Results Stated in Chapter II

Since a result called the Projection Theorem is fundamental to the first three proofs, it will be presented at the outset. [71]

The Projection Theorem establishes an important relationship between projection data of a function f and its Fourier transform $\hat{f}$ given by:

II.1 $$\hat{f}(\xi) = (2\pi)^{\frac{1}{2}} \int_{R^2} f(z) \exp(-i <\xi,z>) dz.$$

Now if I.2 is recalled:

$$P_\theta f(v) = \int_{-\infty}^{+\infty} f(v + t\vec{\theta}) \, dt \quad \text{for } v \text{ in } \theta^\perp, \text{ then}$$

II.2

$$\int_{\theta^\perp} P_\theta f(v) \rho(<v,u>) dv = \int_{R^2} f(w) \rho(<w,u>) dw, \quad \text{for } u \text{ in } \theta^\perp$$

follows, where $<s,z>$ represents the scalar product of the vectors s and z in R^2 and $\rho(t)$ is any function of a single variable, since when u is in $\theta^\perp$

$$\int_{R^2} f(w)\rho(<w,u>)dw = \int_{\theta^1}\int_{-\infty}^{+\infty} f(v+t\vec{\theta})\rho(<v+t\vec{\theta},u>)dt\,dv$$

$$= \int_{\theta^1} \rho(<v,u>)\left(\int_{-\infty}^{+\infty} f(v+t\vec{\theta})dt\right)dv.$$

(Since $u \in \theta^1$, $<v+t\vec{\theta},u> = <u,v>$.)

When $\rho(t) = \exp(-it)$ and powers of 2π are taken into account, II.2 becomes:

Theorem The Projection Theorem

II.3 $(P_\theta f)\widehat{\,}(\xi) = (2\pi)^{\frac{1}{2}} \hat{f}(\xi)$ for ξ in $\widehat{\theta^1}$, where $\widehat{\theta^1}$ is the topological dual of θ^1.

(Of course, θ^1 and $\widehat{\theta^1}$ are just isomorphic to copies of the real line.)

Proof of Proposition 0

It is sufficient to show that if $f + g \in L_0^2(R^2)$ with the same projection data as f then $g = 0$. In that case, $P_\theta g = 0$ for all $\vec{\theta}$ in an infinite subset of S^1. Therefore, $(P_\theta g)\widehat{\,} = 0$ for all $\vec{\theta}$ in an infinite subset of S^1. By the projection theorem g vanishes on an infinite set of lines through the origin. Since $g \in L_0^2(R^2)$, its Fourier transform is a real analytic function on R^2. It is well known that real analytic functions cannot vanish on an infinite set of lines through the origin without vanishing identically. □

Proof of Theorem (K.T. Smith)

This theorem depends on two results. The projection theorem has already been proved.

The second theorem needed is on the solvability of partial differential equations due to B. Malgrange and L. Ehrenpreis. Let $z \in R^k$ for some positive integer k. Let $q(z)$ be a polynomial in $z_1, z_2, \ldots, z_k$ and let $Q(D)$ be the linear differential operator obtained by replacing z_j by $D_j = -i\partial/\partial z_j$. $Q(D)$ may be written as:

$$Q(D) = \sum_\alpha c_\alpha D^\alpha \quad \text{where, for} \quad \alpha = (\alpha_1, \alpha_2, \ldots, \alpha_k)$$

$$D^\alpha = D_1^{\alpha_1} D_2^{\alpha_2} \ldots D_k^{\alpha_k} .$$

Definition: By a fundamental solution of $Q(D)$ we mean a distribution E on R^k such that $Q(D) E = \delta$. Where δ is the distribution satisfying $\delta(z) = 0$ if $z \neq 0$,

$$\int_{R^k} \delta \, dz = 1, \text{ and } f*\delta = \delta*f = f \text{ for any } f \in C_0^\infty(R^k).$$

The importance of the notion of fundamental solution is due to the fact that $u = E * f$, where $f \in C_0(R^k)$, gives a solution to $Q(D)u = f$. In fact, differentiation of such a convolution product provides:

$$Q(D)u = (Q(D)E) * f = \delta * f = f.$$

Theorem (B. Malgrange and L. Ehrenpreis. Proved independently.) [83]

There exists a fundamental solution for every linear partial differential equation with constant coefficients. [83, p. 182]

Proof of Theorem 1 (K. Smith):

The idea of the proof is to find an infinitely differentiable $g = f - f'$ with the property that $P_{\vec{\theta}_j} g = 0$ for some previously chosen set of projection angles, $\{\vec{\theta}_j\}_{j=1}^m$. The polynomial,

$$q(z) = <z, \vec{\theta}_1><z, \vec{\theta}_2> \cdots <z, \vec{\theta}_m>$$

vanishes on each line $\theta_j^\perp$. Let Q be the corresponding differentiable operator defined by replacing z_k by $-i\partial/\partial z_k$. Let L be the compact subset of the interior of the support of f where we wish to arbitrarily specify f'. Let $h_0 = f'$ on L, i.e., h_0 is an arbitrary infinitely differentiable function defined on a neighborhood on L. Let $h_1 = f - h_0$. Let ϕ be any C^∞ function which is 1 on a neighborhood of L and 0 outside a slightly larger neighborhood which is itself contained in support(h_1) $\cap$ support(f). By the theorem of B. Malgrange on the solvability of linear partial differential equations with constant coefficients we can solve $Q(h_2) = h_1$. We will now show that $g = Q(\phi(z) \cdot h_2(z))$ works. The only thing which is not clear is: $P_{\theta_j} g = 0$ for all $j = 1, 2, \ldots, m$. But by the Projection Theorem

$$(P_\theta g)\hat{\,}(\xi) = (2\pi)^{\frac{1}{2}} \hat{g}(\xi) =$$
$$(2\pi)^{\frac{1}{2}} q(\xi)(\phi(\xi) \cdot h_2(\xi))\hat{\,} \text{ for } \xi \in \theta^{\perp}$$

Since q is zero on $\theta_j^{\perp}$, so is $P_{\theta_j} g$, for all $j = 1, 2, \ldots m$. □

Proof of Proposition II.1:

(a) Since each P_{θ_j} is linear, {ghost w.r.t. $\{\theta_j\}$} is a linear subspace. Therefore,

{ghost w.r.t. $\{\theta_j\}_{j=1}^k$} = $\bigcap_{j=1}^k$ {ghost w.r.t. $\{\theta_j\}$} is a linear subspace, so it must be closed under linear combinations.

(b) The fact that $g(z) - g(z + t_0 \vec{\theta}_{k+1})$ is an element of {ghost w.r.t. $\{\theta_j\}_{j=1}^k$} is an application of (a). Therefore, it is enough to show that

$$P_{\theta_{k+1}}(g(z) - g(z + t_0 \vec{\theta}_{k+1})) = 0 .$$

$$P_{\theta_{k+1}}(g(z) - g(z + t_0 \vec{\theta}_{k+1}))(v) \quad \text{for} \quad v \in \theta_{k+1}^\perp$$

$$= \int_{-\infty}^{+\infty} g(v + t\vec{\theta}_{k+1})dt - \int_{-\infty}^{+\infty} g(v + t_0\vec{\theta}_{k+1} + t\vec{\theta}_{k+1}) \, dt$$

$$= \int_{-\infty}^{+\infty} g(v + t\vec{\theta}_{k+1})dt - \int_{-\infty}^{+\infty} g(v + t\vec{\theta}_{k+1})dt = 0 .$$

(c) It is enough to show that $(P_{\theta_j}(p * g))^\wedge = 0$ since if so then $P_{\theta_j}(p * g) = 0$. But, $(P_{\theta_j}(p * g))\hat{}(\xi) = (2\pi)^{\frac{1}{2}} (p * g)^\wedge(\xi)$ for ξ in $\theta_j^\perp$, by the Projection Theorem. Therefore, $(P_{\theta_j}(p * g))^\wedge(\xi) = (2\pi)^{\frac{1}{2}} \hat{p}(\xi)\hat{g}(\xi) = \hat{p}(\xi)(P_{\theta_j} g)^\wedge(\xi) = p(\xi) \cdot 0 = 0$.

(d) This is an easy consequence of the linearity properties of the definition of P_θ. □

Proof of Proposition II.2

We will show that $K \in N$ by using mathematical induction on j with the following inductive step:

h_{j-1} is a ghost w.r.t. $\{\theta_i\}_{i=1}^{j-1}$ implies

h_j is a ghost w.r.t. $\{\theta_i\}_{i=1}^{j}$. In fact, this is easy to prove from part (b) of Proposition II.1 since it implies that

(i) h_1 is a ghost w.r.t. $\{\theta_i\}$ when $k = 0$, and

(ii) h_j is a ghost w.r.t. $\{\theta_i\}_{i=1}^{j}$ when $k = j-1$ and we know that h_{j-1} is a ghost w.r.t. $\{\theta_i\}_{i=1}^{j-1}$.

□

Proof of the Corollary to Proposition II.2

Let X be the vector space containing f. By the hypotheses of the corollary we get that $P_{\theta_j}(f) = P_{\theta_j}(f+K)$ for all $j \leq m$. Therefore, f and $f + K$ are indistinguishable from their projection data alone.

□

CHAPTER III

A RECONSTRUCTION SPACE WHICH DOES NOT CONTAIN THE
OBJECTIVE FUNCTION

The idea which transforms the problem of reconstruction from projections into a solvable problem is the use of a reconstruction space which is different from the space within which the objective function, f , lies. For practical purposes it will be sufficient to reconstruct f (or find an approximation to f) with finite resolution. Any picture in the real world has finite resolution, anyway.

Of course, as soon as it is decided to find a finite dimensional reconstruction, h , a new problem arises: how should h be defined? This question has been answered in various ways. [62] The most prominent reconstruction spaces are described in the next two sections.

A Reconstruction Space based on the Fourier Transform

In their paper [20] of 1970, Crowther, De Rosier, and Klug presented their choice of a reconstruction space. Their choice was a natural consequence of the fact that they were using the (transmission) electron microscope. Instead of using the projection data directly they first Fourier-transformed their data and began to deduce the structure of the Fourier transform of f by a simple application of the Projection Theorem (p. 31). This technique is natural to an electron microscope since it has been shown [30] that it is possible to directly obtain the Fourier-transformed projection data from the microscope. The reconstruction space they used simply required the determination of a finite number of Fourier coefficients of f.

The use of reconstruction spaces that depend on Fourier techniques is common in the literature. Crowther, DeRosier and Klug's choice is justified by use of Shannon's Sampling theorem, which **says that a function,** k, which has a compactly supported Fourier transform is completely determined by k's values on a countable set of equally spaced points [69].

The general Fourier technique chooses a finite dimensional reconstruction space of the form

III.1

$$\hat{Z}(N) = \left\{ k(z) = \sum_{j=1}^{N} a_j \exp(-2\pi i <z,z_j>) : z_j \in R^2, a_j \in C \right\}$$

Since f has compact support, Shannon's Sampling Theorem can be applied to $\hat{f}$, the Fourier transform of f:

$$\hat{f} \text{ is determined by } \left\{ \hat{f}(z_{jk}) : z_{jk} = (j,k) \in Z^2 \right\}.$$

(This follows from the fact that, for $z_{jk} = (j,k)$ we have

$$f = \sum_{j,k=-\infty}^{\infty} <f, \exp(i(jx+ky))> \exp(i(jx+ky)) \text{ , where}$$

$$\hat{f}(z_{jk}) = <f, \exp(i(jx+ky))>$$

$$= \int_{R^2} f(s) \exp(i <z_{jk}, s>) \, ds \, .) \qquad [69]$$

Now we can choose $\hat{Z}(N)$ such that the special subset of R^2 for sampling corresponds to Shannon's Sampling Theorem, i.e.,

III.2

$$\hat{Z}(N) = \left\{ k(z) = \sum_{j=1}^{N} a_j \exp(-2\pi i <z, z_j>) : a_j \in C \text{ , and} \right.$$

$$\left. \{z_j\}_{j=1}^{N} = \{(x,y): x,y \in Z \text{ and } |x|, |y| \le n\} \right\},$$

note that $N = (2n + 1)^2$ if n is an integer.

The value of the reconstruction depends, among many other things, on the fact that not all of the sampling required by Shannon's theorem is used to find the reconstruction. This expresses the finite character of the reconstruction.

A further emphasis of the idea of using a finite dimensional reconstruction is demonstrated by many examples in our environment. The following list of frequently utilized reconstruction spaces are similar in that they can be described mathematically by the same model.

Description of our choice of the Reconstruction Space

The reconstruction space used in this study is approximately the same space which is utilized by (black and white) television, newsprint, satellite telemetry, and the human eye for image production. In each case, a square is subdivided into n^2 smaller, uniform squares of side L/n where L is the length of the side of the entire picture. An image is produced on a television screen by varying the brightness from small square to small square within the picture. Images in a newspaper consist of a lattice ($n \times n$) of points which are used as the center of a small black disk; brightness is achieved by varying the diameter of each of these disks. A satellite transmits a photograph by assigning a positive number to each of n^2 points; each such number represents brightness. The eye is arranged so that its rods (and cones) are in an array with a relatively constant distance between adjacent light receptors; the brightness perceived is transmitted to the brain (in parallel) in just the same way as the satellite transmits its information (in series). This is the same space that is used with the EMI, and this space is the most common choice of researchers who study reconstruction from projections.

Definition of the reconstruction space, $Z(n)$

(i) Let n be a particular positive integer.

(ii) Subdivide $I^2 = \{(x,y) \in R^2 : 0 \leq x,y \leq 1\}$ into n^2 uniform smaller squares by drawing the $n-1$ horizontal line segments $\{(x, \frac{j}{n}) \in R^2 : 0 \leq x \leq 1\}_{j=1}^{n-1}$ and the $n-1$ vertical line segments $\{(\frac{i}{n}, y) \in R^2 : 0 \leq y \leq 1\}_{i=1}^{n-1}$.
Call $_nI^2_{ij}$ the subsquare of I^2 of side $1/n$ with upper right vertex at $\frac{1}{n}(i,j) \in R^2$.

(iii) The *characteristic function of* $_nI^2_{ij}$, denoted by $\chi_{_nI^2_{ij}}$, is identically one on $_nI^2_{ij}$ and zero elsewhere.

(iv) $Z(n)$ is the set of functions defined on I^2 which are constant on each $_nI^2_{ij}$, i.e.,

$$Z(n) \equiv \left\{ k(z) = \sum_{i,j=1}^{n} a_{ij} \chi_{_nI^2_{ij}}(z) : a_{ij} \in R \right\}.$$

This reconstruction space is frequently described by a checkerboard type diagram of small squares. Determining a particular reconstruction is equivalent to choosing a real number for each of these squares. These small squares have their own name; they are called *pixels*.

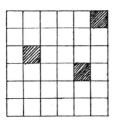

For notational convenience, Z(n) will always denote this particular reconstruction space. That is, Z(n) is the space of functions that are constant on each pixel in a square (n×n) array.

Resolution of a Reconstruction ≡ Picture Resolution

Resolution of an image or an observation is generally defined as the size of the smallest part which can be reliably discriminated. This definition is very easy to understand when applied to Z(n), since the smallest object in this reconstruction space is $\chi_{n^2 I_{ij}}$, i.e., the size of a pixel. Therefore the *resolution of a reconstruction* in Z(n) is 1/n. For clarity we call this quantity the *picture resolution*.

(Kennan Smith has defined resolution of a reconstruction in a somewhat more practical way. He recognizes that a reconstruction $h \in Z(n)$ cannot exactly represent f, moreover because some error occurs in the calculation of h, it should be even more evident that the information about f which can be inferred from h must have poorer resolution than 1/n. For such reasons he prefers to describe the resolution of a reconstruction more conservatively at 2/n. [71])

CHAPTER IV

A MATRIX REPRESENTATION OF THE PROBLEM

Throughout this and the following three chapters, the objective function, f, is assumed to be an element of the finite dimensional reconstruction space, $Z(n)$, for some known n. Under this condition the problem of reconstruction from projections is considered. The question of approximating an arbitrary objective function by an element of $Z(n)$ is discussed in Chapters VIII and IX.

It is possible to express the relationship between $P_\theta f$ and f in the form of a matrix equation. The necessary justification is given after the matrix equation is presented.

Consider the following definitions:
1) Let A be a $(n^2 \times 1)$ column vector which represents f as an element of $Z(n)$;
2) let $v_1, v_2, \ldots, v_k, v_{k+1}$ be the images of vertices of pixels projected onto $\theta^\perp$ and numbered sequentially along $\theta^\perp$;*
3) let W_θ be a $(k \times 1)$ column vector such that its ith component $(W_\theta)_i$ is equal to the integral of the projection data from v_i to v_{i+1} ; and
4) let U_θ be the $(k \times n^2)$ matrix which satisfies the equation

$$U_\theta A = W_\theta$$.

---
*k depends on n and θ.

U_θ is called the observational matrix for P_θ as it acts on A.

For example: Let $n = 2$, $\theta = 45°$ and $f \in Z(2)$ given by

$$f = \sum_{i,j=1}^{n} a_{ij} \, X_{nI_{ij}^2}.$$

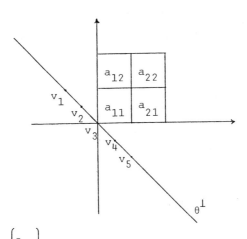

Let $A \equiv \begin{pmatrix} a_{12} \\ a_{11} \\ a_{22} \\ a_{21} \end{pmatrix}$ be a (4×1) vector which represents f as element of $Z(n)$.

Notice that although there are nine vertices of pixels in this reconstruction space (located at the points: (0,0), (½,0), (1,0), (0,½), (½,½), (1,½), (0,1), (½,1) and (1,1)), the lines through these vertices parallel to $\vec{\theta}$ only intersect θ^1 in five points: $v_1 = (-\frac{1}{2}, \frac{1}{2})$, $v_2 = (-\frac{1}{4}, \frac{1}{4})$, $v_3 = (0,0)$, $v_4 = (\frac{1}{4}, -\frac{1}{4})$ and $v_5 = (\frac{1}{2}, -\frac{1}{2})$. As a result, k + 1, which is the number or projected images of vertices of pixels, must be five. Therefore, $W_\theta = W_{45^\circ}$ is a (4×1) vector, i.e.,

$$W_\theta \equiv \begin{pmatrix} (W_\theta)_1 \\ (W_\theta)_2 \\ (W_\theta)_3 \\ (W_\theta)_4 \end{pmatrix}.$$

Its ith component is given by

$$(W_\theta)_i \equiv \int_{v_i}^{v_{i+1}} P_\theta f(s) \, ds . \qquad \text{In fact,}$$

$$(W_\theta)_1 \equiv \int_{v_1}^{v_2} P_\theta f(s) \, ds$$

$$\equiv \int_{v_1}^{v_2} \int_{-\infty}^{+\infty} \sum_{i,j=1}^{2} a_{ij} \, X_{I_{ij}^2}(s + t\vec{\theta}) \, dt \, ds$$

$$= (1/2) \, a_{12} . \qquad \text{Similarly,}$$

$$(W_\theta)_2 \equiv \int_{v_2}^{v_3} P_\theta f(s) \, ds$$

$$= \int_{v_2}^{v_3} \int_{-\infty}^{+\infty} \sum_{i,j=1}^{2} a_{ij} \, X_{I_{ij}^2}(s + t\vec{\theta}) \, dt \, ds$$

$$= (1/2)(a_{12} + a_{11} + a_{22}), \text{ and etc.}$$

Let U_θ be the observational matrix for P_θ as it acts on A, i.e., U_θ satisfies $U_\theta A = W_\theta$, then we must have

$$U_\theta = \begin{pmatrix} \frac{1}{2} & 0 & 0 & 0 \\ \frac{1}{2} & \frac{1}{2} & \frac{1}{2} & 0 \\ 0 & \frac{1}{2} & \frac{1}{2} & \frac{1}{2} \\ 0 & 0 & 0 & \frac{1}{2} \end{pmatrix}.$$

It will be shown that all of the information in $P_\theta f$ is expressed by $U_\theta A = W_\theta$ in the sense that the vector W_θ determines the function $P_\theta f$. (This fact is fundamental to the discussion about the required resolution in the projection data, in Chapter V.)

Definition: The *projected image of vertices of pixels in the direction* $\vec{\theta}$, for a fixed n, is the set of points in $\theta^\perp$, $\{v_1, v_2, \ldots, v_{k+1}\}$, such that for each $i = 1, 2, \ldots, k+1$, the line $\{v_i + t\vec{\theta} : t \text{ in } R\}$ contains a vertex of a pixel of $Z(n)$.

We shall see that all of the information in $P_\theta f$ is available once
{average value of $P_\theta f$ on the interval (v_i, v_{i+1}):
for $i = 1, 2, \ldots, k\}$
is known.

To begin the explanation of the information in $P_\theta f$, we will first examine $P_\theta \chi_{n^I_{ij}}^2$, i.e., we shall consider the projection of the function which is constantly 1 on the i,j^{th} pixel and zero elsewhere.

The diagrams of the next page indicate how $P_{\{\theta\}}\big|_{Z(n)}$ depends on θ. The upper right diagram shows θ between $45°$ and $90°$; the line, $\theta^\perp$, is also indicated. The upper left diagram shows dotted line segments parallel to $\vec{\theta}$. The two lower diagrams are the corresponding figures when θ is between $0°$ and $45°$.

By considering $P_\theta \chi_{n^I_{ij}}^2$ in this way we see that the region where this is non-zero has end points L and M. It is also clear that on the region with end points 0 and K $P_\theta \chi_{n^I_{ij}}^2$ is constant. The dotted lines, therefore, indicate where the integration changes its local dependence on the parameter t. It is an easy calculation to show:

$$P_\theta \bar{X}_{n^I_{ij}}^2(t) = \begin{cases} \frac{1/n - t/\sin\theta}{\cos\theta} & \text{for } \frac{(\sin\theta-\cos\theta)}{n} \leq t \leq (\sin)/n \\ 1/(n\sin\theta) & \text{for } 0 \leq t \leq \frac{(\sin\theta-\cos\theta)}{n} \\ \frac{1/n + t/\sin\theta}{\cos\theta} & \text{for } -(\cos\theta)/n \leq t \leq 0 \\ 0 & \text{otherwise} \end{cases}$$

for θ in $[45°, 90°)$ and

$$P_\theta X_{n^I_{ij}}^2(t) = \begin{cases} \frac{1/n - t/\sin\theta}{\cos\theta} & \text{for } 0 \leq t \leq (\sin\theta)/n \\ 1/(n\cos\theta) & \text{for } (\sin\theta-\cos\theta)/n \leq t \leq 0 \\ \frac{1/n + t/\cos\theta}{\sin\theta} & \text{for } -\frac{\cos\theta}{n} \leq t \leq \frac{\sin\theta-\cos\theta}{n} \\ 0 & \text{otherwise .} \end{cases}$$

for θ in $(0°, 45°]$

49

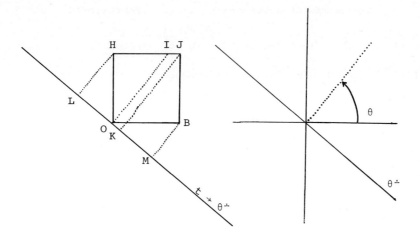

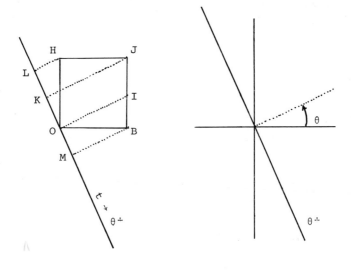

The expression for $P_\theta \chi_{{}_n I_{ij}^2}(t)$ demonstrates that there are only three regions within θ^1 where $P_\theta \chi_{{}_n I_{ij}^2}(t)$ is non-zero and has distinctive dependence on t. Those regions have end points K, L, M and O in the previous diagrams; the set {K, L, M and O} is exactly the set of projected images of vertices of the pixel ${}_n I_{ij}^2$. If we let $\vec{\theta}^1$ be the point on θ^1 where $t = 1$, for some θ between 45° and 90°, then $M = [(\sin\theta)/n]\vec{\theta}^1$, $K = [(\sin\theta - \cos\theta)/n]\vec{\theta}^1$, and $L = [-(\cos\theta)/n]\vec{\theta}^1$, and $P_\theta \chi_{{}_n I_{ij}^2}(t)$ has a **simple** formula on each of the following sub-intervals of θ^1:
(L,O), (O,K), and (K,M).

Definition: Let $F(x) = y$ be a real-valued function of a real variable. F is called *affine linear* if $F(x) = ax + b$ for some real constants a and b.

Proposition IV.1 $P_\theta \chi_{{}_n I_{ij}^2}(t)$ is a continuous function of t when θ is not a multiple of 90°. For any $\vec{\theta} \in S^1$, $P_\theta \chi_{{}_n I_{ij}^2}(t)$ is affine linear on each of the intervals: (v_1, v_2), (v_2, v_3) and (v_3, v_4) where $\{v_1, v_2, v_3 \text{ and } v_4\}$ is the set of projected images of the vertices of the pixel ${}_n I_{ij}^2$.

Now we are ready to consider the general situation, i.e., consider $P_\theta f$ for arbitrary f in $Z(n)$ and arbitrary $\vec{\theta} \in S^1$.

Proposition IV.2: Let $v_1, v_2, \ldots, v_k, v_{k+1}$ be the projected images of vertices of pixels in the direction $\vec{\theta}$, then f in $Z(n)$ has a projection, $P_\theta f$, which is completely determined by the k real numbers in the set:

{average value of $P_\theta f$ on the inverval (v_i, v_{i+1}):

for $i = 1, 2, \ldots, k$}*

Finally, the problem of reconstruction from finitely many projections (when f is considered to be an element of $Z(n)$) is:

determine f from $\{P_{\theta_1} f, P_{\theta_2} f, \ldots, P_{\theta_m} f\}$.

If we let

$$U = \begin{pmatrix} U_{\theta_1} \\ U_{\theta_2} \\ \vdots \\ U_{\theta_m} \end{pmatrix} \quad \text{and} \quad W = \begin{pmatrix} W_{\theta_1} \\ W_{\theta_2} \\ \vdots \\ W_{\theta_m} \end{pmatrix},$$

then the entire task of reconstruction from projections can be expressed by:

$$\boxed{U A = W}.$$

---
*The idea which generated this proposition grew out of a personal communication with Ron Huesman, Lawrence Radiation Laboratory, Berkeley.

Proofs of Propositions Stated in Chapter IV

Proof of Proposition IV.1: This follows from the formuli on page 48. □

Proof of Proposition IV.2:

For $i \leq k$, $P_\theta f: [v_i, v_{i+1}] \to R$ can be represented by
$$P_\theta f(s) = P_\theta f(v_i) + s(P_\theta f(v_{i+1}) - P_\theta f(v_i)) \text{, for}$$
$s \in [v_i, v_{i+1}] \subseteq \theta^1$, since $P_\theta \chi_{n^I{}^2_{ij}}(\cdot)$ is an affine linear function on $[v_i, v_{i+1}]$ by Proposition IV.3 and since $P_\theta f(s)$ is a linear combination of such functions. Therefore, $P_\theta f: \theta^1 \to R$ is determined by its values on $\{v_1, v_2, \ldots, v_k, v_{k+1}\}$. But $P_\theta f(v_1) = P_\theta f(v_{k+1}) = 0$, so it is enough to find $\{P_\theta f(v_2), \ldots P_\theta f(v_k)\}$. Furthermore:

IV.1 $P_\theta f(v_{i+1}) = 2(\text{average value of } P_\theta f \text{ on } [v_i, v_{i+1}]) - P_\theta f(v_i)$

since $P_\theta f(s)$ is affine linear on $[v_i, v_{i+1}]$ its average value occurs at the midpoint of (v_i, v_{i+1}), therefore

average value of $P_\theta f$ on $[v_i, v_{i+1}]$ $= P_\theta f(\frac{1}{2}(v_i + v_{i+1}))$

$$= \tfrac{1}{2} P_\theta f(v_i) + \tfrac{1}{2} P_\theta f(v_{i+1}) \text{ .}$$

However, since we have $P_\theta f(v_1) = 0$ and by hypothesis we have $\{\text{average value of } P_\theta f \text{ on } [v_i, v_{i+1}]: \text{ for } i = 1, 2, \ldots, k-1\}$, can apply IV.1 k-1 times and we are done. □

CHAPTER V

RESOLUTION IN THE PROJECTION DATA

In practice, projection data only contains a finite amount of information. For this reason experimenters have introduced a notion of resolution in any type of observational data.

We now assume that the projection data are known. The projection data are given at a finite number of points. These points cannot be any closer than some minimal distance that depends on the nature of the way the observations are made. This distance, w, we call the *resolution of the projection data*.

Projection Angles Affect the Required Resolution

Proposition IV.2 tells us that if we require all of the information in $P_\theta f$, then we need only determine the average value of $P_\theta f$ on each of the subsets of $\theta^\perp$ with endpoints within the set of projected images of vertices of pixels in the direction $\vec{\theta}$. It is, therefore, necessary that

$$w \leq \min(||v_i - v_{i+1}||)$$

in order that all of the information in $P_\theta f$ be retrievable in practice, where $\{v_1, v_2, \ldots, v_{k+1}\}$ is the set of projected images of vertices of pixels in the direction $\vec{\theta}$.*

The next proposition shows that it is possible to determine $\min(||v_i - v_{i+1}||)$ when n and θ are given.

Proposition V.1 Let n be any positive integer, $\vec{\theta}$ any element of S^1, and $\{v_1, v_2, \ldots, v_{k+1}\}$ be the set of projected images of vertices of pixels in the direction $\vec{\theta}$, then

$$n[\min(||v_i - v_{i+1}||)] = (\tan^2\theta + 1)^{-1/2} \min|s(\tan\theta) - t|$$

---

*Since $\min(||v_i - v_{i+1}||)$ is actually a function of theta, the projection angle, we shall frequently use $R(\theta)$ to represent $\min(||v_i - v_{i+1}||)$ in the following.

where s and t range over integers, $0 \leq s, t \leq n$ such that $s(\tan\theta) - t \neq 0$.

The following table utilizes this proposition for various choices of n and θ. The values of s and t where $\min|s(\tan\theta)-t|$ is attained is also given. $\theta = 38.5°$ is used to demonstrate that for some θ, $\min(\|v_i - v_{i+1}\|)$ is not proportional to $1/n$.

Application of Proposition V.1

n	θ in degrees	$\min\|s(\tan\theta)-t\|$	attained at s	t	$\min(\|v_i - v_{i+1}\|)$	Picture Resolution
2	00.0	1.000	0	1	.500	.500
2	45.0	.707	1	0	.353	.500
2	26.6	.447	1	1	.283	.500
2	38.5	.017	1	1	.008	.500
3	00.0	1.000	0	1	.333	.333
3	45.0	.707	1	0	.236	.333
3	26.6	.447	1	1	.188	.333
3	18.4	.316	1	0	.106	.333
3	33.7	.277	1	1	.094	.333
3	38.5	.017	1	1	.006	.333
4	00.0	1.000	0	1	.250	.250
4	45.0	.707	1	0	.177	.250
4	26.6	.447	1	1	.141	.250
4	18.4	.316	1	0	.079	.250
4	33.7	.277	1	1	.070	.250
4	14.0	.245	1	0	.056	.250
4	38.5	.017	1	1	.0043	.250
5	38.5	.011	5	4	.0022	.200

Farey Series and Projection Angles

In the course of this study of the problem of Reconstruction from Projections (which is strongly focused on the use of a specific reconstruction space, namely $Z(n)$, the *pixel reconstruction space*) two new and somewhat independent results point to the significance of a special subset of S^1 from which the projection angles should be chosen. These results (Theorems 1 and 2) appear in this and the next chapter.

This special subset of S^1 was first recognized by Mersereau and Oppenheim for its seemingly unsatisfactory property of being the only angles in S^1 which are not clearly capable of providing reliable reconstructions from very accurate data. [55] (See page 63.) The second use of this subset of S^1 was made by H. R. P. Ferguson in his paper "Reconstruction of plane objects by Farey dissection of Radon's integral solution" presented among the post-deadline papers at the conference on Image Processing at Stanford in August, 1975. [26] Ferguson recognized that a convenient description of these interesting angles could be made using the concept of Farey Series.

Definition : A Farey series F_n of order n is the set of fractions h/k with $0 \leq h \leq k$, h and k relatively prime, $1 \leq k \leq n$, and arranged in ascending order of magnitude.

For example : F_5 is 0/1, 1/5, 1/4, 1/3, 2/5, 1/2, 3/5, 2/3, 3/4, 4/5, 1/1.

A most interesting and surprisingly useful fact stands as a theorem in a well known book on number theory [5].

Theorem : If h/k and h'/k' are two successive fractions in F_n, then kh' - hk' = 1.

This theorem, the last proposition and the following simple definition establishes an important result concerning projection angles taken from the set:

Definition : $\bar{\theta}(n) = \{\vec{\theta}$ in $S^1 |\tan\theta, \mathrm{ctn}\theta, -\tan\theta$ or $-\mathrm{ctn}\theta$ in $F_n\}$.
The significance of $\bar{\theta}(n)$ is: $\vec{\theta}$ in $\bar{\theta}(n)$ if and only if there is a line with slope = $\tan\theta$ which connects two vertices of pixels of $Z(n)$.

For example : $\bar{\theta}(2) = \{0°, 90°, 45°, 135°, \arctan(2), \arctan(-2), \arctan(1/2), \arctan(-1/2)\}$. On the other hand when $\theta = 38.5°$, $\vec{\theta}$ is not in $\bar{\theta}(n)$ for $n = 1, 2, \ldots, 10$ since $\theta = \arctan(.799)$.

Proposition V.2 shows that when θ has a rational tangent, $\min(||v_i - v_{i+1}||)$ is proportional to 1/n when n is sufficiently large.

Proposition V.2 (Ferguson) : Let $\vec{\theta}$ in $\bar{\theta}(n)$, n a positive integer, and $\{v_1, v_2, \ldots, v_{k+1}\}$ be the set of projected images of vertices of pixels then

$$n [\min(||v_i - v_{i+1}||)] = (p^2 + q^2)^{-1/2},$$

where $\tan\theta = p/q$, an element of F_n.

Application of Proposition V.2

$\tan\theta$	approximate value of θ in degrees	Ratio of Required Data Resolution to chosen Picture Resolution
0/1	00.000000	1.000
1/1	45.000000	.707
1/2	26.565057	.447
1/3	18.434952	.316
2/3	33.690073	.277
1/4	14.036247	.242
3/4	36.869904	.200
1/5	11.309935	.196
2/5	21.801413	.185
3/5	30.963762	.171
1/6	9.462324	.164
4/5	38.659815	.156
1/7	8.130104	.141
2/7	15.945399	.137
3/7	23.198594	.131
5/6	39.805578	.128

Significance of the Farey Projection Angles

This discussion has indicated that the angles which are not in $\bar{\theta}(n)$ are not as useful for solving the practical problem of reconstruction from projections. Notice that the distance between projected images of vertices of pixels decrease more rapidly for $\theta = 38.5°$ than the corresponding numbers for $\theta = 45°$ or $\theta = \arctan(1/2)$ while n increases. (See the table following Proposition V.1) In fact, the ratio of the required resolution remained constant for $\theta = 45°$, $\arctan(1/2)$, $\arctan(1/3)$, and $\arctan(2/3)$ as Proposition V.2 predicts, but the ratio decreased for $\theta = 38.5°$ when n changed from 4 to 5. Although these observations strongly suggest the preferred use of projection angles chosen from $\overline{\theta}(n)$ over those chosen outside of $\overline{\theta}(n)$, only the following theorem lays to rest any remaining doubt concerning the advantages obtained by picking projection angles from within $\overline{\theta}(n)$.

Theorem 1: Let n be fixed, let $(\phi,\psi) \subseteq S^1$ such that $(\phi,\psi) \cap \overline{\theta}(n)$ is not empty, and let $v_1(\theta), v_2(\theta), \ldots, v_{k(\theta)+1}(\theta)$ be the projected images of vertices of pixels in the direction $\vec{\theta}$, then

$$R(\theta) = \min(\| v_1(\theta) - v_{i+1}(\theta) \|)$$

attains its maximum on (ϕ,ψ) with the set $(\phi,\psi) \cap \overline{\theta}(n)$.[*]

In other words, within any small range of perspective projection angles (namely (ϕ,ψ)) which intersects $\overline{\theta}(n)$, the relevant information about the determination of a reconstruction is most accessible if that projection angle is taken from $\overline{\theta}(n)$. *Since resolution in the projection data is often a limiting factor, and because the angles in $\overline{\theta}(n)$ require the least resolution, angles in $\overline{\theta}(n)$ are the most useful for practical applications.*

Proofs of Results Stated in Chapter V

Because the two propositions are simply corollaries of the lemmas needed for the proof of Theorem 1 it would be most efficient to present those lemmas and simply indicate when the two propositions are easily established.

Lemma V.1 : Let n be any arbitrary positive integer, $\vec{\theta}$ any element of S^1, then a vertex of a pixel in the reconstruction space $Z(n)$ looks like $[1/n](s,t)$ where $0 \leq s,t \leq n$ and s and t are integers. The distance between projected images of $[1/n](s_1, t_2)$ is given by:

$$[1/n[[(s_1-s_2)^2 + (t_1-t_2)^2]^{1/2} |\sin\{\theta - \arctan[(t_1-t_2)/(s_1-s_2)]\}|.$$

---

[*] I am indebted to Professor Phillip Leonard, Arizona State University, for a key observation which lead to the proof of this result.

Proof of Lemma V.1

    Draw a right triangle formed by:

  i)   connect $[1/n](s_1,t_1)$ to $[1/n](s_2,t_2)$ to form the hypotenuse,

  ii)   draw a line parallel to $\vec{\theta}$ and passing through $[1/n](s_1,t_1)$, and

  iii) draw a third line through $[1/n](s_2,t_2)$ and perpendicular to $\vec{\theta}$.

    Because the second line segment of this triangle is parallel to $\vec{\theta}$, the distance from the third vertex of the triangle to $[1/n](s_2,t_2)$ must be the same as the distance between the projected images of $[1/n](s_1,t_1)$ and $[1/n](s_2,t_2)$ in the direction $\vec{\theta}$.

    The angle at $[1/n](s_1,t_1)$ must be $\pm[\theta - \arctan\{(t_1-t_2)/s_1-s_2)\}]$. Now use the definition of the sine of this angle to get the desired formula.

Lemma V.2: If θ in $(0°, 45°)$ then $R(\theta)$ derives its value at one of the two displacements involving $(0,0)$.

Proof of Lemma V.2

    Let $[1/n](s_1,t_1)$ and $[1/n](s_2,t_2)$ be any pair of vertices of pixels whose projected images are as close together as any such pair. Assume that $s_1 \geq s_2$.

$$s_1 > s_2 \text{ implies that } t_1 \geq t_2$$

since if not then $\sin[\theta - \arctan\{(t_2 - t_1)/(s_2 - s_1)\}] > \sin\theta$ and the projected distance of $[1/n](1,0)$ to $(0,0)$ is smaller than the corresponding distance for the pair $[1/n](s_1,t_1)$, $[1/n](s_2,t_2)$ by Lemma V.1. Thus $[1/n](s_1-s_2,t_1-t_2)$ is a vertex of a pixel for $Z(n)$.

Lemma V.3: Let (s,t) and $y = mx$ by any point in the plane and any line through the origin. Then the distance from (s,t) to $y = mx$ is given by $|ms - t|(m^2 + 1)^{-1/2}$.

Proof of Lemma V.3   Exercise in analytic geometry.

Corollary to Lemma V.3   If $\theta = \arctan(p/q)$ then the projected image of (s,t) is $|ps-tq|(p^2 + q^2)^{-1/2}$ units from $(0,0)$.

Proof of Proposition V.1

    By Lemma V.2, $\min(\|v_i - v_{i+1}\|)$ derives its value at one of the displacements involving $(0,0)$. Therefore, we need only consider the distances between the origin and the projected images of vertices of pixels, $\{[1/n](s,t) | 0 \leq s,t, \leq n\}$. Notice that $Z(n)$ is symmetric about the lines $y = x$ and $y = 1 - x$; therefore, the restriction to θ in $(0°, 45°)$ is artificial. Now apply Lemma V.3 with $m = \tan\theta$.

Lemma V.4: Let p and q be relatively positive prime integers with $0 \leq p/q \leq 1$ and $q \leq n$ then $nR(\arctan(p/q)) = (p^2 + q^2)^{-1/2}$.

Proof of Lemma V.4

Let $\theta = \arctan(p/q)$. By Lemma V.2 and V.3 and its corollary,

$$n R(\theta) = \min [|ps - tq| (p^2 + q^2)^{-1/2}] > 0$$

where s and t range over all non-negative integers $\leq n$. Since the numerator is always a positive integer, we are done if for some vertex of a pixel, $[1/n](k,h)$, we get $|pk-qh| = 1$, but this is satisfied by the vertex defined by choosing an h/k which is consecutive with p/q in some Farey series.

Proof of Proposition V.2

This is merely a corollary of Lemma V.4 since the restriction on θ imposed by $0 < p/q \leq 1$ is artificial -- as explained in the proof of Proposition V.1 above.

Lemma V.5: Let $\theta_1 = \arctan(p_1/q_1)$, $\theta_2 = \arctan(p_2/q_2)$ for $0 \leq \theta_1 < \theta_2 \leq 45°$ be two consecutive discontinuities of $R(\theta)$ for some fixed n then for $\theta \in (\theta_1, \theta_2)$,

$$n(\tan^2\theta+1)^{1/2} R(\theta) = \min(|q_1(\tan\theta)-p_1|, |q_2(\tan\theta)-p_2|).$$

Lemma V.5 requires two sublemmas.

Sublemma V.5.1: The function, $\min\{|q_i(\tan\theta)-p_i| [\tan^2\theta+1]^{-1/2}\}$ over $i = 1$ and 2, reaches its maximum in (θ_1, θ_2) at $\arctan[(p_1+p_2)/(q_1+q_2)]$.

Proof of Sublemma V.5.1

Since the two functions are strictly monotonic in opposite senses we can set them equal and solve for θ.

Corollary to Sublemma V.5.1 (q_1, p_1) is closer to $y = (\tan\theta)x$ than is (q_2, p_2) as long as $\arctan(p_1/q_1) \leq \theta < \arctan[(p_1+p_2)/(q_1+q_2)]$.

Sublemma V.5.2 : If s, t, h, k, p, and q are positive numbers and $t/s \leq h/k < p/q$ then $ps - qt > pk - qh$.

Proof of Sublemma V. 5.2

$t/s \leq h/k$ implies $tk \leq hs$ which implies $(t-h)k \leq (s-k)h$. Since $p/h > q/k$, we get: $(s-k)p > (t-h)q$ which implies $ps - qt > pk - qh$.

Proof of Lemma V.5

By Lemmas V.2 and V.3,
$$n(\tan^2\theta+1)^{1/2} R(\theta) = \min(|s(\tan\theta) - t|, |S(\tan\theta) - T|)$$
for some (s,t) and (S,T) which may depend on θ. (Notice that t/s and T/S are in F_n.) We will show that (s,t), (S,T) is constantly (q_1, p_1), (q_2, p_2) for θ in (θ_1, θ_2). By the Corollary to Sublemma V.5.1 it is enough to show that (s,t) is further from $y = (\tan\theta)x$ than is (q_1, p_1) whenever $t/s < p_1/q_1$, similarily for $T/S > p_2/q_2$. Let h/k be the element of the Farey Series of order n immediately before p_1/q_1, then:

distance from (s,t) to $y = (\tan\theta)x$
$\geq$ distance from (s,t) to $y = (p_1/q_1)x$
$= |p_1 s - q_1 t|(p_1^2 + q_1^2)^{-1/2}$
$\geq |p_1 k - q_1 h|(p_1^2 + q_1^2)^{-1/2}$
$= (p_1^2 + q_1^2)^{-1/2}$
$> [(p_1 + p_2)^2 + (q_1 + q_2)^2]^{-1/2}$
$= (p_2 q_1 - q_2 p_1)[(p_1 + p_2)^2 + (q_1 + q_2)^2]^{-1/2}$
$= [(p_1 + p_2)q_1 - (q_1 + q_2)p_1][(p_1+p_2)^2 + (q_1+q_2)^2]^{-1/2}$
$=$ distance from (q_1, p_1) to $y = [(p_1 + p_2)/(q_1 + q_2)]x$
$\geq$ distance from (q_1, p_1) to $y = (\tan\theta)x$.

Corollary to Lemma V.5 Let $\theta_1 = \arctan(p_1/q_1)$ and $\theta_2 = \arctan(p_2, q_2)$ be two consecutive discontinuities of $R(\theta)$ for fixed n. Then $R(\theta)$ reaches its maximum on (θ_1, θ_2) at $\theta = \arctan[(p_1+p_2)/(q_1+q_2)]$ where R takes the value $[1/n][(p_1+p_2)^2 + (q_1+q_2)^2]^{-1/2}$.

Proof of Corollary to Lemma V.5

The fact that $R(\theta)$ reaches its maximum at

$$\theta = \arctan[(p_1 + p_2)/(q_1 + q_2)]$$

is the application of Sublemma V.5.1 and Lemma V.5. The content of the fifth to last to the second to last lines of the proof of Lemma V.5 shows that

$$n[(p_1+p_2)^2+(q_1+q_2)^2]^{1/2} \, R(\arctan[p_1+p_2)/(q_1+q_2)]) =$$
$$\min(|p_1+p_2)s-(q_1+q_2)t|) \quad \text{attains its value for the } (s,t) = (q_1,p_1).$$

The significance of this fact is explained in beginning of the proof of Lemma V.4.

Proof of Theorem 1

We will see that it is sufficient to prove the theorem when $(\phi,\psi) \cap \overline{\theta}(n) = \{\theta_1, \theta_2\}$. First, we show that $R(\theta)$ attains its maximum on $[\theta_1, \theta_2]$ at either θ_1 or θ_2. It is sufficent to prove this when $0° \leq \theta_1 < \theta_2 \leq 45°$ since $R(\theta) = R(\theta + 90°) = R(-\theta) = R(45° - \theta)$ because of the symmetries of the reconstruction space. By Lemma V.4, $R(\theta_1) = R(\arctan[p_1/q_1]) = (p_1^2 + q_1^2)^{-1/2}$, $R(\theta_2) = (p_2^2 + q_2^2)^{-1/2}$. By the Corollary to Lemma V.5,

$$\max_{\theta \text{ in } (\theta_1,\theta_2)} R(\theta) = [(p_1 + p_2)^2 + (q_1 + q_2)^2]^{-1/2}.$$

Therefore,

$$\max_{\theta \text{ in } [\theta_1,\theta_2]} R(\theta) \quad \text{occurs at either } \theta_1 \text{ or } \theta_2.$$

For the more general situation either $(\phi,\psi) \cap \overline{\theta}(n)$ has (case a) one element, or (case b) two or more elements. Case a can be handled in light of the last two expressions involving max $R(\theta)$ on (θ_1,θ_2) and and $[\theta_1,\theta_2]$ respectively. Clearly, if θ_3 is the discontinuity of $R(\theta)$ immediately preceding θ_1 then the maximum value of $R(\theta)$ on (θ_3,θ_2) occurs at θ_1. Case b can be handled by separately considering each pair of consecutive elements of $(\phi,\psi) \cap \overline{\theta}(n)$.

CHAPTER VI

RESULTS ESTABLISHING THE UNIQUENESS OF A RECONSTRUCTION

Throughout this chapter we will assume that the objective function, f, is an element of the finite dimensional space $Z(n)$ for some known n and that under these conditions we would like to know when f is determined by a finite set of projections.

Since the transformation, P_θ, is now defined on and restricted to $Z(n)$, there is a problem equivalent to the problem of uniquely determining f, namely: when does $P_{\{\theta\}}$ have a trivial nullspace?[*] This equivalence of problems is expressed in the following proposition.

PROPOSITION VI.1 $\{P_{\theta_1} f, P_{\theta_2} f, \ldots, P_{\theta_m} f\} \equiv P_{\{\theta\}} f$ determines $f \in Z(n)$ if and only if

$$N \equiv \bigcap_{j=1}^{m} \text{nullspace of } P_{\theta_j}\Big|_{Z(n)} \text{ is } \{0\}.$$

This chapter will answer the question of unique determination of f from its projection data by fully characterizing N.

---

[*] The equivalence of the two problems does not depend on the fact that P_θ is restricted to $Z(n)$. This is evident in the proof of Proposition VI.1.

At this point, the reader will remember that θ represents an angle measured counter-clockwise from the x-axis and that $\vec{\theta}$ represents the corresponding element of S^1. ($S^1 = \{z \in R^2 : |z| = 1\}$.) f is the *objective function* which we have assumed to be an element of $Z(n)$, the *reconstruction space* defined on *pixels*. A *projection of* f *in the direction* $\vec{\theta}$ is

$$P_\theta f(v) = \int_{-\infty}^{+\infty} f(v + t\vec{\theta}) dt, \quad \text{for} \quad v \in \theta^\perp$$

(as explained in Chapter I, p. 7). $P_{\{\theta\}}$ can be thought of as a linear transformation from the space containing f to the space containing the projection data (as explained in Chapter II, p. 25).

The first result which begins to characterize the nullspace of $P_{\{\theta\}}$ was established by Oppenheim and Mersereau [59]. Later Kennan Smith [71] showed that it could be generalized to all finite dimensional reconstruction spaces.

Proposition VI.2 (Mersereau and Oppenheim) Let n be a fixed positive integer, and let $\vec{\theta}$ be in S^1 such that $\vec{\theta} \notin \bar{\theta}(n)$, then $P_\theta h \equiv 0$ implies that $h \equiv 0$, for $h \in Z(n)$.

The significance of the hypothesis that $\vec{\theta} \notin \bar{\theta}(n)$ arises from the fact that this condition guarantees that any line passing through one vertex of a pixel can not pass through any other vertex -- this insures that the images of two pixels cannot exactly overlap each other in $P_\theta f$.

When $\vec{\theta} \in \bar{\theta}(n)$, P_θ usually does not have a trivial nullspace. In addition, more than one projection angle is normally used for collecting data: therefore, we are now interested in the set of g's in $Z(n)$ such that $P_{\theta_j}(g) = 0$ for $j = 1, 2, \ldots, m$. We shall use the terminology $P_{\{\theta\}} \equiv P_{\{\theta_j\}} \equiv P_{\{\theta_j\}_{j=1}^m}$ to represent the set of projection transformations that have been used to produce the observational data. When the term, *nullspace of* $P_{\{\theta\}}$ is used, we are referring to the set of g in $Z(n)$ such that $P_{\theta_j}(g) = 0$, for $j = 1, 2, \ldots, m$. (To remind the reader of the terminology, we call such a g a *ghost with respect to* $\{\theta_j\}_{j=1}^m$.)

The following theorem explains more than what is required; not only does it explain when $P_{\{\theta\}}$ will have a non-trivial nullspace, but also it says exactly what a ghost must be.

Theorem 2 Let $\{\vec{\theta}_j\}_{j=1}^m \subseteq \{\arctan \frac{p}{q} : p, q \in Z\} \subseteq S^1$ and let $g \in Z(n)$ for some $n \in Z^+$ which satisfies $P_{\theta_j}(g) = 0$, for $j = 1, 2, \ldots, m$. Then there exist $L \in Z^+$, $\{b_i\}_{i=1}^L \subseteq R$, $\{z_i\}_{i=1}^L \subseteq R^2$ and $K \in Z(n)$ such that

$$g(z) = \sum_{i=1}^L b_i K(z - z_i), \text{ where } K \text{ is defined by}$$

i) $\tan\theta_j = p_j/q_j$ for p_j and q_j relatively prime integers for each j,

ii) $h_0 = \chi_{n^{-1}i_0 j_0}$ for some i_0 and $j_0 \leq n$,

iii) $h_j(z) = h_{j-1}(z) - h_{j-1}(z - \frac{1}{n}(q_j, p_j))$,   and

iv) $K = h_m$.

Furthermore $N \neq \{0\}$ if an only if

$$n \geq 1 + \max\left(\sum_{j=1}^{m} |p_j|, \sum_{j=1}^{m} |q_j|\right).*$$

(A complete explanation of what Theorem 2 says follows a discussion of an interpretation of Proposition VI.2 and Theorem 2.)

Interpretation of the Two Uniqueness Results:
Proposition VI.2 and Theorem 2

The proper interpretation of these results belongs in two different settings. For the theoretician, it is desirable to ponder these results in the case where each projection, $P_{\theta_j} f$, is noiseless and can be read at every point along $\theta_j^\perp$. However, for the practical scientist, it is necessary to recognize that observational projection data is subject to noise and can only be sampled a finite number

---
*
The definition of K here is slightly different from the one presented in Proposition II.2.

of times, i.e., the resolution of the projection data is limited. In either case, a researcher would ask about the available picture resolution (of the reconstruction).

When a projection is considered to be perfect, i.e., it exactly corresponds to $P_\theta f(z) = \int_{-\infty}^{+\infty} f(z + t\vec{\theta})dt$, for z in $\theta^\perp$, then we are assuming that we have noiseless data which can be read at infinitely many places. In this impractical situation, Oppenheim and Mersereau's result says that any single projection of f can be used to completely determine f. Their result (Proposition VI.2) places no restriction on n where n^2 is the dimension of $Z(n)$, the reconstruction space. As a consequence, their result says that no matter how large n is, if n is known and $P_\theta f$ is given for $\tan\theta$ irrational, then f is determined, since all angles in $\bar{\theta}(n)$ have rational tangents.

We must be careful in the interpretation of Proposition VI.2 with respect to the issue of picture resolution. It may seem to suggest that infinitely small picture resolution is possible, but on the contrary, Proposition VI.2 depends on knowing exactly how much picture resolution exists, i.e., knowing n. In contrast to K. Smith's Theorem where an infinite amount of detail is allowed, Oppenheim and Mersereau's result makes no attempt to find details of f smaller than a predetermined pixel size.

It is interesting to see the very dramatic difference between the cases $\vec{\theta} \in \bar{\theta}(n)$ and $\vec{\theta} \notin \bar{\theta}(n)$. When the angle is chosen perfectly, the picture resolution of an f which can be determined by $P_\theta f$ seems to have no lower limit. But when the angle is only slightly (imperceivably slightly) changed, $\tan\theta$ can become rational and Theorem 2 predicts a bound on the picture resolution. In fact, if $\tan\theta = \frac{p}{q}$ for p, q relatively prime integers, then Theorem 2 implies that when f is in $Z(n)$ for $n \geq \max(1 + |p|, 1 + |q|)$ then $P_\theta f$ does not determine f.

Now if we restrict our attention only to $\vec{\theta}$ in S^1 such that $\tan\theta$ is rational, then an interesting phenomenon occurs. Let f_n be in $Z(n)$ and $\theta_n = \arctan(1/n)$, Theorem 2 then says that f_n is determined by $P_{\theta_n} f_n$. Therefore, careful choice of the projection angle, even when restricted to $\{\vec{\theta} \in S^1: \tan\theta \text{ is rational}\}$, allows for unlimited picture resolution.

It is therefore clear that any important distinction between the meaning of these results must exist in another context.

There is in Practice a Limitation on the Resolution in $P_\theta f$

For empirical reasons, we will assume that there is a number w which specifies the smallest region in θ^1 where $P_\theta f$ should be sampled. (See Chapter V, p. 53.) Certainly w can vary from experiment to experiment, but there is always some limit on the fineness of the observations.

(Recall that w is called the *resolution of the projection data*.)

It is now possible to make a critical distinction between the results: Proposition VI.2 and Theorem 2. Proposition IV.2 from the fourth chapter tells us that if we desire all of the information in $P_\theta f$ then we need only determine the average value of $P_\theta f$ on each of the subsets of θ^1 with end-points among the set of projected images of vertices of pixels in the direction $\vec{\theta}$. Proposition V.2 says that when $\vec{\theta} \in \bar{\theta}(n)$, it is easy to determine the minimal distance between adjacent projected images of vertices. As was emphasized in Chapter V **by Theorem 1, the angles in** $\bar{\theta}(n)$ **are** of more practical use than those outside of $\bar{\theta}(n)$ since they require the least resolution in the projection data. *Proposition VI.2 sacrifices the resolution in the projection data in favor of the picture resolution (of the reconstruction)*. But Proposition V.2 says that the ratio of resolution in the projection data to the resolution in the reconstruction can be kept within practical limits when $\vec{\theta} \in \bar{\theta}(n)$. Therefore, *Theorem 2* which specifies the resolution in the reconstruction when $\{\vec{\theta}_j\}_{j=1}^m \subseteq \bar{\theta}(n)$, *is practically important since it only depends on the availability of a reasonable amount of resolution in the projection data*. For that reason, Oppenheim and Mersereau's result is of less practical significance than Theorem 2.

In the next chapter, a statistical treatment utilizes Theorem 2 and deals with the recognition that

Figure IV.0

-50	-50	-50	-40	20	30	30	28	14	-50	-50	-50
-50	-50	-20	30	13	3	3	13	30	15	-50	-50
-50	-40	30	8	3	3	3	3	3	30	-40	-50
-50	15	15	3	3	3	3	3	3	18	15	-50
-50	30	3	3	3	3	3	3	3	5	30	-50
-50	30	3	3	3	3	3	3	3	3	30	-50
-50	15	13	3	3	3	3	3	3	13	20	-50
-50	5	30	6	3	3	3	3	6	30	-40	-50
-50	-50	10	30	6	3	3	7	30	-20	-50	-50
-50	-50	-50	10	30	30	30	30	-20	-50	-50	-50
-50	-50	-50	-50	-50	-50	-50	-50	-50	-50	-50	-50

-50	-50	-50	-50	-50	-50	-50	-50	-50	-50	-50	-50
-50	-50	-50	-40	20	31	29	28	14	-50	-50	-50
-50	-50	-20	29	13	3	3	13	31	15	-50	-50
-50	-40	31	8	3	4	2	3	3	29	-40	-50
-50	15	15	3	3	3	3	3	3	18	15	-50
-50	29	3	2	3	3	3	3	4	5	31	-50
-50	31	3	4	3	3	3	3	2	3	29	-50
-50	15	13	3	3	3	3	3	3	13	20	-50
-50	5	29	6	3	4	2	3	6	31	-40	-50
-50	-50	10	31	6	3	3	7	29	-20	-50	-50
-50	-50	-50	10	30	29	31	30	-20	-50	-50	-50
-50	-50	-50	-50	-50	-50	-50	-50	-50	-50	-50	-50

Although these two images are diagnostically different, they could not be distinguished from their projection data. This is an example of insufficient design. The two simulated reconstructions are displayed on a 12 by 12 array. (This represents an array of uniform squares, each over a centimeter on a side.) Each number is chosen to be approximately one tenth of the appropriate EMI number for slices taken through a patient's head. If projections of these two images are taken at exactly eight equally spaced projection angles, then the corresponding projection data will not disagree more than one part in one thousand at any datum. Notice that the lower image shows diagnostically significant fluctuations which are not present in the upper image; in particular, consider the 4's in the intracranial portion of the array.

projection data is frequently noisy.

Explanation of Theorem 2

A consequence of Theorem 2 is that $N \equiv$ nullspace of $P_{\{\theta\}} \equiv \bigcap_{j=1}^{m}$ nullspace $P_{\theta_j}\big|_{Z(n)}$ is composed of elements of the form:

$$\sum_{i=1}^{p} a_i K(z - z_i), \quad \text{where} \quad a_i \in R \text{ and } z_i \in R^2.$$

That is, N is generated as a vector space by some subset of $\{K(z - z_i) \mid z_i \in R^2\}$. Since $N \subseteq Z(n)$, it must be that $K(z - z_i) \in Z(n)$ for some $z_i \in R^2$, as well as, $K(z) \in Z(n)$. We can therefore learn all about N by first examining K.

Remember that the definition of K depends on the choice of $\{\theta_1, \theta_2, \ldots \theta_m\}$. It is recursively defined by

(i) $h_0 = X_{n^{I^2}i_0 j_0}$ for some i_0, j_0 positive integers $\leq n$;

(ii) $h_j(z) = h_{j-1}(z) - h_{j-1}(z - \frac{1}{n}(q_j, p_j))$, where

$\tan\theta_j = \frac{p_j}{q_j}$ for p_j and q_j relatively prime integers; and

(iii) $K = h_m$.

(Although this definition does not uniquely define K, since i_0, j_0 are not specified, this arbitrary aspect of the definition of K is reflected in the statement of the theorem, since $K(z - z_i)$ is used.)

On the next page, a series of four simple examples of K are demonstrated. The fifth diagram shows a more general element of N.

Figure VI.1 demonstrates K when $\{\theta_1, \theta_2, \ldots, \theta_m\} = \{0°\}$. (For convenience, $\boxed{+1}$ represents $\chi_{_n I_{ij}^2}$ and $\boxed{-1}$ represents $-\chi_{_n I_{ij}^2}$ on the respective pixels.)

Figure VI.2 demonstrates K when $\{\theta_1, \theta_2, \ldots, \theta_m\} = \{0°, 90°\}$.

Figure VI.3 demonstrates K when $\{\theta_1, \theta_2, \ldots, \theta_m\} = \{0°, 90°, \arctan 3\}$.

Figure VI.4 demonstrates K when $\{\theta_1, \theta_2, \ldots, \theta_m\} = \{0°, 90°, \arctan 3, \arctan \frac{-1}{5}\}$.

The choice of this sequence of figures was no accident; it was made to demonstrate how K can be obtained from its recursive definition. Suppose we would like to obtain K by using its definition when n is greater than 7, m = 4 and $\{\theta_1, \theta_2, \ldots, \theta_m\} = \{0°, 90°, \arctan 3, \arctan \frac{-1}{5}\}$, we can now describe each of the intermediate recursively defined functions:

Let $h_0 = \chi_{_n I_{i_0 j_0}^2}$ where $_n I_{i_0 j_0}^2$ is the pixel marked $\boxed{+1}$ in Figure VI.1. Then h_1 is drawn in Figure VI.1 since

$$h_1(z) = h_0(z) - h_0(z - \frac{1}{n}(q_1, p_1)),$$

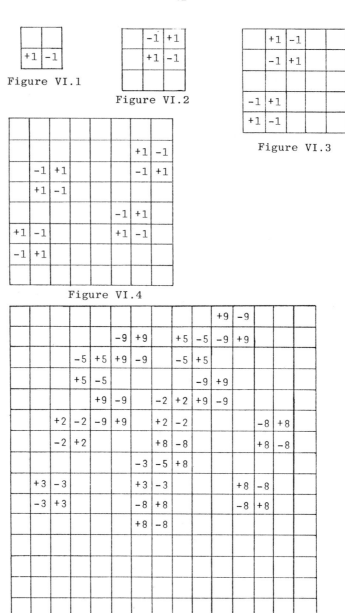

Figure VI.1

Figure VI.2

Figure VI.3

Figure VI.4

Figure VI.5

but $\tan\theta_1 = \tan 0° = \frac{0}{1}$. Therefore $p_1 = 0$, $q_1 = 1$, and and $h_0 = \chi_{nI^2_{i_0 j_0}}$, and we get

$$h_1(z) = \chi_{nI^2_{i_0 j_0}}(z) - \chi_{nI^2_{i_0 j_0}}(z - \tfrac{1}{n}(1,0)).$$

Then $h_2(z)$ is drawn in Figure VI.2, since

$$h_2(z) = h_1(z) - h_1(z - \tfrac{1}{n}(q_2, p_2)) = h_1(z) - h_1(z - \tfrac{1}{n}(0,1))$$

$$= \chi_{nI^2_{i_0 j_0}}(z) - \chi_{nI^2_{i_0 j_0}}(z - \tfrac{1}{n}(1,0))$$

$$- \chi_{nI^2_{i_0 j_0}}(z - \tfrac{1}{n}(0,1)) + \chi_{nI^2_{i_0 j_0}}(z - \tfrac{1}{n}(1,1)).$$

Similarly, $h_3(z)$ is drawn in Figure VI.3, and of course, $h_4(z) = K$ is drawn in Figure VI.4.

The theorem speaks of general elements of N; it says that they are represented by

$$\sum_{i=1}^{P} a_i K(z - z_i), \quad \text{where} \quad a_i \in R, \; z_i \in R^2.$$

The next figure demonstrates that when n is much greater than

$$1 + \max\left(\sum_{j=1}^{m} |p_j|, \sum_{j=1}^{m} |q_j|\right)$$

very complicated ghosts are possible. Let
$\{\theta_1, \theta_2, \ldots, \theta_m\} = \{0°, 90°, \arctan 3, \arctan \frac{-1}{5}\}$ (i.e., we have chosen the same projection angles which were used to determine K in Figure VI.4). Figure VI.5 is a diagram of a ghost with respect to $\{\theta_j\}_{j=1}^{4}$. In this case n was chosen to be 15 and the element of the nullspace drawn is given by

$$g(z) = 9 K(z) + 5 K(z - \frac{1}{15}(-2,-1))$$

$$+ 3 K(z - \frac{1}{15}(-3,-4)) + -8 K(z - \frac{1}{15}(2,-5)) .$$

The Theorem says that if N is not equal to {0} then

VI.1 $\quad n \geq 1 + \max \left(\sum_{j=1}^{m} |p_j| , \sum_{j=1}^{m} |q_j| \right).$

This follows from the fact that if $g \in N$ and $g \neq 0$ then K must exist as an element of $Z(n)$. But K requires $\frac{1}{n}(1 + |q_1| + |q_2| + \ldots + |q_m|)$ units in the x-direction and $\frac{1}{n}(1 + |p_1| + |p_2| + \ldots + |p_m|)$ units in the y-direction. Therefore if $g \neq 0$ then $K \neq 0$ and the above formula holds. In fact, this is the key to its practical applications, since

$f \in Z(n)$ is determined by $P_{\{\theta\}} f$

$$iff \quad N \equiv nullspace \text{ of } P_{\{\theta\}} = \{0\}$$

$$iff \quad n < 1 + max\left(\sum_{j=1}^{m} |q_j|, \sum_{j=1}^{m} |p_j|\right).$$

Uniquely Determined Picture Resolution

We will recall that the resolution of a reconstruction can be defined as the size of the smallest pixel. For our needs this is adequately represented by $1/n$. A reconstruction, h, is of no value if it is not uniquely determined by its projection data, since any element, g, in N can be added to such an h to get h + g which has exactly the same projection data. Therefore, the term *picture resolution* is only meaningful when it refers to a reconstruction that was *uniquely determined*.

Since the previous discussion linked $\{\theta_1, \theta_2, \ldots, \theta_m\}$, the set of projection angles, to n such that when $f \in Z(n)$, $P_{\{\theta\}} f$ *uniquely determines* f, we can now link $\{\theta_1, \theta_2, \ldots, \theta_m\}$ to *uniquely determined picture resolution*.

The next table presents uniquely determined picture resolution along side $\{\theta_1, \theta_2, \ldots, \theta_m\}$.

$\{\theta_1, \theta_2, \ldots, \theta_m\}$	$\{\tan \theta_j : j=1,2,\ldots,m\}$	Uniquely determined picture resolution
$0°$	0	$1/1$
$0°, 90°$	$0, \infty$	$1/1$
$0°, 90°, 45°$	$0, \infty, 1$	$1/2$
$36.85° = \arctan \frac{3}{4}$	$3/4$	$1/4$
$5.7° = \arctan \frac{1}{10}$	$1/10$	$1/10$
$62.3° = \arctan 1.9$	$19/10$	$1/19$
$0°, \ldots, \arctan 2, \arctan \frac{-1}{2}$	$0, \infty, 1, -1, \frac{1}{2}, 2, -\frac{1}{2}$	$1/8$
$44.35° = \arctan .99$	$99/100$	$1/100$

This table says that when projection data is collected in the directions $\vec{\theta}_1, \vec{\theta}_2, \ldots, \vec{\theta}_m$ then $f \in Z(n)$ is uniquely determined by $P_{\{\theta\}}f$, when $1/n$ is listed under the column entitled "uniquely determined picture resolution". For example: On the seventh line of the table $\{\theta_1, \theta_2, \ldots, \theta_m\} = \{0°, 90°, 45°, 135°, \arctan \frac{1}{2}, \arctan 2, \arctan \frac{-1}{2}\}$, when $f \in Z(8)$, $P_{\{\theta\}}f$ determines f. (This is true since $\{\theta_1, \theta_2, \ldots, \theta_m\}$ specifies that $\{(q_j, p_j) \in Z^2 : \tan \theta_j = \frac{p_j}{q_j}$ for p_j and q_j relatively prime $\}$ equals $\{(1,0),(0,1),(1,1),(-1,1),(2,1),(1,2),(-2,1)\}$. Therefore, $\sum_j |q_j| = 1 + 0 + 1 + 1 + 2 + 1 + 2 = 8$.)

It is well to compare these facts with some similar results by Crowther, DeRosier, and Klug [20]. Although they used a completely different argument to arrive at their results, they present a formula which expresses picture resolution (of a reconstruction) in terms of m equally spaces projections:

$$\text{Resolution} \cong \pi/m \quad \text{(for large enough } m\text{)}.$$

It is not possible to directly compare their result with the consequences of Theorem 2 because m equally spaced angles $\theta_1, \theta_2, \ldots, \theta_m$ would usually involve at least one θ_j such that $\tan\theta_j$ is irrational. (This is true, since for most values of m, $\tan(\pi/m)$ is irrational.) However, when $\{\theta_1, \theta_2, \ldots, \theta_m\}$ is chosen so as to get the least desirable picture resolution according to Theorem 2 some kind of comparison can be made.

Picture Resolution as a Function of m Projection Angles

Crowther, De Rosier and Klug's resolution is a function of m equally spaced projection angles; Resolution $\cong \frac{\pi}{m}$.

Theorem 2 can be used to pick m different projection angles so that the worst uniquely determined resolution is obtained;

$$\text{Resolution} = \left[\max \left(\sum_{j=1}^{m} |q_j| \, , \, \sum_{j=1}^{m} |p_j| \right) \right]^{-1} .$$

(Let ϕ be the angle between adjacent projection angles when the projection angles are chosen according to C., De R. and K, i.e., $m\phi = 180°$. The set of projection angles which are chosen according to Theorem 2 just accumulates a single new element for each increase in m ; therefore, that set can be described by a cumulative list: let θ_m be that new element.)

m	ϕ	θ_m	Dimensions of K	Picture Resolution C.,DeR.&K.	Picture Resolution Theorem 2
1	180°	0°	2 X 1	1/1	1/1
2	90°	90°	2 X 2	1/1	1/1
3	60°	45°	3 X 3	1/1	1/2
4	45°	135°	4 X 4	1/1	1/3
5	36°	arctan $\frac{1}{2}$	6 X 5	1/1	1/5
6	30°	arctan $\frac{2}{1}$	7 X 7	1/2	1/6
7	25.7°	arctan $-\frac{1}{2}$	9 X 8	1/2	1/8
8	22.2°	arctan $-\frac{2}{1}$	10 X 10	1/2	1/9
9	20°	arctan $\frac{1}{3}$	13 X 11	1/2	1/12
10	18°	arctan $\frac{3}{1}$	14 X 14	1/3	1/13
11	16.4°	arctan $-\frac{1}{3}$	17 X 15	1/3	1/16
12	15°	arctan $-\frac{3}{1}$	18 X 18	1/4	1/17
13	13.8°	arctan $\frac{2}{3}$	21 X 20	1/4	1/20
14	12.8°	arctan $\frac{3}{2}$	23 X 23	1/4	1/22
15	12°	arctan $-\frac{2}{3}$	26 X 25	1/5	1/25
16	11.2°	arctan $-\frac{3}{2}$	28 X 28	1/5	1/27
17	10.6°	arctan $\frac{1}{4}$	32 X 29	1/5	1/31
18	10°	arctan $\frac{4}{1}$	33 X 33	1/5	1/32
19	9.5°	arctan $-\frac{1}{4}$	37 X 34	1/6	1/36
20	9°	arctan $-\frac{4}{1}$	38 X 38	1/6	1/37
21	8.6°	arctan $\frac{3}{4}$	42 X 41	1/6	1/41
22	8.2°	arctan $\frac{4}{3}$	45 X 45	1/7	1/44
23	7.8°	arctan $-\frac{3}{4}$	49 X 48	1/7	1/48
24	7.5°	arctan $-\frac{4}{3}$	52 X 52	1/7	1/51
25	7.2°	arctan $\frac{1}{5}$	57 X 53	1/8	1/56
26	6.9°	arctan $\frac{5}{1}$	58 X 58	1/8	1/57
27	6.7°	arctan $-\frac{1}{5}$	63 X 59	1/8	1/62
28	6.4°	arctan $-\frac{5}{1}$	64 X 64	1/9	1/63
29	6.2°	arctan $\frac{2}{5}$	69 X 66	1/9	1/68
30	6°	arctan $\frac{5}{2}$	71 X 71	1/9	1/70
31	5.8°	arctan $-\frac{2}{5}$	76 X 73	1/10	1/75
32	5.6°	arctan $-\frac{5}{2}$	78 X 78	1/10	1/77
33	5.5°	arctan $\frac{3}{5}$	83 X 81	1/10	1/82
34	5.3°	arctan $\frac{5}{3}$	86 X 86	1/11	1/85
35	5.1°	arctan $-\frac{3}{5}$	91 X 89	1/11	1/90
36	5°	arctan $-\frac{5}{3}$	94 X 94	1/11	1/93

It should be clear that Theorem 2 predicts much better picture resolution than does the formula of Crowther, DeRosier and Klug. In fact, over the range of m used in the last table, when Crowther, DeRosier and Klug gets π/m, Theorem 2 gets 1/ℓ where:

$$(\pi/m)^2 \approx 1/\ell .$$

That is, Theorem 2 predicts picture resolution which is approximately the square of the picture resolution predicted by Crowther, DeRosier, and Klug.

Finally, an interesting result occurs when m = 18 and $\{\theta_1, \theta_2, \ldots, \theta_m\}$ is again chosen according to Theorem 2 but with another idea in mind. In an attempt to approximate the angles specified by Crowther, DeRosier, and Klug, let the projection angles be the arctan of the following rational numbers: $\frac{0}{1}, \frac{1}{6}, \frac{1}{3}, \frac{5}{6}, \frac{6}{5}, \frac{5}{3}, \frac{1}{3}, \frac{6}{1}, \frac{1}{0}, \frac{-6}{1}, \frac{-1}{3}, \frac{-5}{3}, \frac{-6}{5}, \frac{-5}{6}, \frac{-1}{3},$ and $\frac{-1}{6}$. When m = 18, C. DeR., and K's work requires the use of {0°, 10°, 20°, ..., 170°}. These correspond to the following angles respectively: 0°, 9.5°, 18.5°, 31°, 39.8°, 50.2°, 59°, 71.5°, 80.5°, 90°, 99.5°, 108.5°, 121°, 129.8°, 140.2°, 149°, 161.5°, and 170.5°. In each case, the chosen angle is within 1.5° of the appropriate angle. Since $\Sigma |p_j| = \Sigma |q_j| = 61$, the dimensions of K are 62 × 62 and reconstruction on a 61 × 61 grid is uniquely determined; $f \in Z(61)$ is uniquely determined by $P_{\{\theta\}}f$. This is a significant improvement over C., De R. and K as their work would predict the use of a 6 × 6 grid since $\frac{1}{6} \approx \pi/18$.

Proofs of Results stated in Chapter VI

Proof of Proposition VI.1

If there exists a g such that $P_{\theta_i} g \equiv 0$ for $i = 1, 2, \ldots, m$ then $P_{\theta_i}(f + g) = P_{\theta_i} f$ so $P_{\{\theta\}}(f + g) = P_{\{\theta\}} f$; therefore, f is not uniquely determined by its projection data.

Suppose whenever $P_{\theta_i} g \equiv 0$ for $i = 1, 2, \ldots, m$ it is also true that $g \equiv 0$, and suppose f and f' have the same projection data, i.e., $P_{\{\theta\}} f = P_{\{\theta\}} f'$. Therefore, $P_{\{\theta\}}(f - f') \equiv 0$, i.e., $P_{\theta_i}(f - f') \equiv 0$ for $i = 1, 2, \ldots, m$ so it must also be true that $f - f' \equiv 0$ and $f = f'$. □

Proof of Proposition VI.2

Because $\vec{\theta} \notin \bar{\theta}(n)$, any line with slope $= \tan\theta$ that passes through one vertex of some $_n I_{ij}^2$ cannot pass through any other such vertex. As a result one can define a one to one correspondence from the set of pixels in $Z(n)$ to the set of projected images of pixels. Thus if $P_\theta g = 0$ on all the regions $\alpha_i \subseteq \theta^\perp$ then $g \equiv 0$. See diagram.

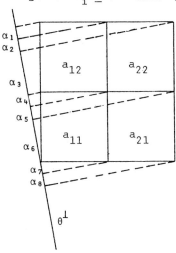

For example:

α_1 determines a_{12}; α_2 and a_{12} determine a_{22}; α_4, a_{12} and a_{22} determine a_{11} etc.

□

The proof of Theorem 2 requires four lemmas.

Lemma 1 Let g be a ghost with respect to $\{\theta_j\}_{j=1}^m$ such that $g \in Z(n)$, then for each θ_j there are two non-collinear line segments with slope $= \tan\theta_j$ which are part of the boundary of the convex hull of the support of g, (denoted by con(supp(g))).

Proof: Pick θ_j and place a ruler on the plane of the support of g, oriented so that its edge makes the angle θ_j with the x-axis. Now slide the ruler along $\theta_j^\perp$ maintaining its orientation until the support of g is the first encountered. There must be two different points in the support of g coincident with the edge of the ruler, otherwise $P_{\theta_j}(g) \neq 0$. □

Notice that the boundary of the con(supp(g)) must also contain line segments parallel to the x-axis simply because $g \in Z(n)$, similarly for the y-axis.

Lemma 2 Let D be a compact convex subset of R^2 with boundary given by a polygon with vertices $\{(x_i, y_i)\}_{i=1}^k \subseteq Z^2 \subseteq R^2$ such that for each $\theta_j \in \{\theta_j\}_{j=1}^m$ there are two edges of the polygon with slope $= \tan\theta_j$, then

$$\max_{i,j} |x_i - x_j| \geq \sum_{j=1}^m |q_j| \quad \text{and} \quad \max_{i,j} |y_i - y_j| \geq \sum_{i=1}^m |p_i|.$$

Proof Let $\emptyset: \{1,2,\ldots,k\} \longrightarrow S^1$, defined by $\emptyset(i) = \arctan((y_{i+1} - y_i)/(x_{i+1} - x_i))$, we shall assume that the vertices are numbered so that as i increases $\emptyset$ maps into S^1 in a counterclockwise manner, we shall also assume that $3\pi/2 \geq \emptyset(k) \geq \pi/2$, $\emptyset(k) = \arctan((y_1 - y_k)/(x_1 - x_k))$.

$$\max_{i,j} |x_i - x_j| = \sum_{\emptyset(i) \geq \pi/2}^{3\pi/2} |x_i - x_{i+1}| \geq \sum_{\theta_j \geq \pi/2}^{3\pi/2} |x_j - x_{j+1}| = \sum_{j=1}^{m} |q_j|.$$

The proof for the y's is similar. □

Lemma 3 Let g be a ghost with respect to $\{\theta_j\}_{j=1}^{m}$ such that $g \in Z(n)$; if $g \neq 0$ then $\text{supp}(g)$ cannot be represented on a rectangular grid of width $\frac{1}{n} \sum_{j=1}^{m} |q_j|$ or height $\frac{1}{n} \sum_{j=1}^{m} |p_j|$. However, if K exists then $V \equiv$ boundary of $\text{con}(\text{supp}(K))$ satisfies:

VI.2 $$n\left(\max_{i,j} |x_i^* - x_j^*|\right) = 1 + \sum_{j=1}^{m} |q_j|$$

and $$n\left(\max_{i,j} |y_i^* - y_j^*|\right) = 1 + \sum_{j=1}^{m} |p_j| ,$$

where $\{(x_i^*, y_i^*)\}$ are the vertices of V.

Proof: As already mentioned, the boundary of the $\text{con}(\text{supp}(g))$ for any $g \in Z(n)$ must have edges parallel to the x and y axes, since, in addition, g is a ghost with respect to $\{\theta_j\}_{j=1}^{m}$ Lemmas 1 and 2 can be applied so that if $\{(x_i, y_i)\}$ are the vertices of the boundary of the convex hull of the support of g then

$$n\left(\max_{i,j} |x_i - x_j|\right) \geq 1 + \sum_{j=1}^{m} |q_j|$$

and $$n\left(\max_{i,j} |y_i - y_j|\right) \geq 1 + \sum_{j=1}^{m} |p_j| .$$

By Proposition II.1 (in Chapter II) it is easy to see that the K defined in the statement of this theorem is a ghost with respect to $\{\theta_i\}_{i=1}^{m}$ whenever it exists. The reader can verify that VI.2 holds. □

At this point we have established all of the practical aspects of the content of Theorem 2. This is true since we now have: If $g \neq 0$ then $n \geq 1 + max\left(\sum_{j=1}^{m} |p_j|, \sum_{j=1}^{m} |q_j|\right)$, which therefore establishes uniquely determined picture resolution.

Lemma 4 Let W be the boundary of the con(supp(g)) for some $g \in Z(n)$ such that g is a ghost wrt $\{\theta_i\}_{i=1}^{m}$. Let $\{W_j\}_{j=1}^{k}$ be the vertices of W. Then for each W_j there exists a $z_j \in R^2$ such that $W_j \in (z_j + V) \subseteq$ con(supp(g)).*

The next five pages provide the proof of Lemma 4. The proof of Theorem 2 is provided on the sixth page following.

---

* V is defined in Lemma 3.

To facilitate the proof of this lemma, we shall need a few notational conventions, definitions, and a few special functions.

Assume that $\emptyset \leq \theta_1 < \ldots < \theta_m \leq \pi + \emptyset$ for some phase angle $\emptyset$. Let $\{V_i\}_{i=1}^{s}$ and $\{W_j\}_{j=1}^{t}$ be the vertices of V and W respectively. (The indices are chosen by counterclockwise rotation along the respective perimeter.) For any vectors A, B, C, D in R^2 some definitions will be made: $\overline{AB}$ is the line segment with end points A and B, $\overleftrightarrow{AB}$ is the line passing through A and B, and $\overrightarrow{V_i} \equiv V_{i+1} - V_i$ and $\overrightarrow{W_j} \equiv W_{j+1} - W_j$.

A useful convention: Whenever indexing is awkward because of the cyclic nature of the vertices of V and W, tacitly assume that the expression is invariant with respect to the **phase angle** $\emptyset$. For example: In the definition of $\overrightarrow{V_i}$ we have tacitly meant that $\overrightarrow{V_s} = V_1 - V_s$.

Definition   Let $G: R^2 \setminus \{0\} \rightarrow S^1 \subseteq R^2$ by $G(C) = C/\|C\|$.

Definition   Let HP: $R^2 \times R^2 \longrightarrow \{\text{closed half-planes} \subseteq R^2\}$ by HP(A,B) = the closed half-plane bounded by $\overleftrightarrow{AB}$ and on your left as you look from A to B.

Definition   For $A, B \in S^1 \subseteq R^2$, "$A \leq B$" means $A = B$ or A is an element of the **complement of** HP(0,B).

Sublemma HP($\cdot$,$\cdot$) has the following properties when $A,B,C,D \in R^2$:

a) $C + D \in HP(A,B) \Leftrightarrow C \in HP(A-D, B-D)$,

b) $C,D \in HP(0,B) \Rightarrow \alpha C + D \in HP(0,B)$, for $\alpha \geq 0$,

c) $C \in HP(A,B) \Leftrightarrow C + \alpha(A-B) \in HP(A,B)$ for α any real number.

d) $G(-B) \leq G(A) \leq G(B) \Leftrightarrow A \in$ interior of $HP(0,B)$.

Sketch of the proof of Lemma 4

The proof proceeds by reducing the problem to Finding z_{j_0} such that $W_{j_0} \in (z_{j_0} + V) \subseteq con(W)$. That result is obtained in stages:

(1) Show that for any W_j there exists a V_{i_j} such that
$$G(\overrightarrow{V_{i_j-1}}) \leq G(\overrightarrow{W_{j-1}}) \quad \text{and} \quad G(\overrightarrow{V_{i_j}}) \geq G(\overrightarrow{W_j}) .$$

(2) Choose $z_{j_0} = W_{j_0} - V_{i_0}$ where V_{i_0} is picked via (1) and W_{j_0} .

(3) Without loss of generality, assume that $j_0 = i_0 = 1$.

(4) $con(W) = \bigcap\limits_{j=1}^{t} HP(W_j, W_{j+1})$.

(5) Pick arbitrary j_* such that $1 \leq j_* \leq t$.

(6) Show that $z_1 + V_{i_*+1} \in HP(W_{j_*}, W_{j_*+1})$, when V_{i_*} is picked via (1) and W_{j_*} .

(7) Show that $(z_1 + V) \subseteq HP(W_{j_*}, W_{j_*+1})$.

(8) Show that we are done since: (2) gives us that $W_{j_0} = W_1 = z_1 + V_1 \in (z_1 + V)$ and (4)-(7) say that $(z_1 + V) \subseteq con(W)$.

Proof of Lemma 4

Since $\operatorname{con}(W) = \operatorname{con}(\operatorname{supp}(g))$; we are done if for each W_{j_0} there exists a z_{j_0} such that $W_{j_0} \in (z_{j_0} + V) \subseteq \operatorname{con}(W)$.

(Step (1):) By Lemma 1 there must be two edges of W which satisfy: $\overrightarrow{W_{j_i} W_{j_i+1}}$ and $\overrightarrow{W_{j_{i+1}} W_{j_{i+1}+1}}$ are parallel to $\vec{\theta}_i$ and $\vec{\theta}_{i+1}$ respectively, and such that $j_i + 1 \leq j_0 \leq j_{i+1}$. This is equally true for the edges of V, but moreover, we also have that $G(\overrightarrow{V_{j-1}}) = \vec{\theta}_i \leftrightarrow G(\overrightarrow{V_j}) = \vec{\theta}_{i+1}$ (unless possibly $G(\overrightarrow{V_j})$ is a multiple of $\pi/2$ when θ_{i+1} is not, in that case $G(\overrightarrow{V_{j+1}}) = \vec{\theta}_{i+1}$). Since the edges of a convex polygon are oriented so that they "turn" in a counterclockwise direction as you travel in a counterclockwise direction along the perimeter, if V_{i_0} is chosen so that $G(\overrightarrow{V_{i_0}}) = \vec{\theta}_{i+1}$, we get: $\vec{\theta}_i = G(\overrightarrow{V_{i_0-1}}) = G(\overrightarrow{W_{j_i}}) \leq G(\overrightarrow{W_{j_0-1}}) < G(\overrightarrow{W_{j_0}}) \leq G(\overrightarrow{W_{j_{i+1}}})$
$= G(\overrightarrow{V_{i_0}}) = \vec{\theta}_{i+1}$. Now choose $z_{j_0} = W_{j_0} - V_{i_0}$. Without loss of generality let $j_0 = i_0 = 1$. The fact that $\operatorname{con}(W) = \bigcap_{j=1}^{t} HP(W_j, W_{j+1})$ is a well known property of convex sets. Therefore, we are done if we can show that $(z_1 + V) \subseteq HP(W_{j_*}, W_{j_*+1})$ for an arbitrary $j_* \leq t$. (Step (6):) Pick V_{i_*} according to the method described in (1) using W_{j_*}. It is easy to check that either $\{\overrightarrow{W_j}\}_{j=1}^{j_*-1} \subseteq HP(0, -\overrightarrow{W_{j_*}})$ or $\{-\overrightarrow{W_j}\}_{j=j_*}^{t} \subseteq HP(0, -\overrightarrow{W_{j_*}})$, use **Sublemma**, d). Since the argument is similar in either case, we will assume the former. Then $W_1 + \sum_{j=1}^{j_*} \overrightarrow{W_j} = W_{j_*+1}$ and
$z_1 + V_1 + \sum_{i=1}^{i_*} \overrightarrow{V_i} = z_1 + V_{i_*+1}$, so

$$W_{j_*+1} - (z_1 + V_{i_*+1}) = \sum_{j=1}^{j_*} \overrightarrow{W_j} - \sum_{i=1}^{i_*} \overrightarrow{V_i}.$$

But the arguments used in Lemmas 1 and 2, as well as the fact that, V_1 and V_{i_*} correspond to W_1 and W_{j_*} via step (1) respectively, imply that for each $i \leq i_*$ there exists an $r_i \geq 1$ and $j_i \leq j_*$ such that

$$r_i \overrightarrow{V_i} = \overrightarrow{W_{j_i}}.$$

We can now apply the various parts of the sublemma and get:

$$\sum_{j=1}^{j_*} \overrightarrow{W_j} - \sum_{i=1}^{i_*} \overrightarrow{V_i} =$$

$$\sum_{1 \leq j \neq j_i} \overrightarrow{W_j} + \sum_{i=1}^{i_*} (\overrightarrow{W_{j_i}} - \overrightarrow{V_i}) \in HP(0, -\overrightarrow{W_{j_*}}).$$

So

$$W_{j_*+1} - (z_1 + V_{i_*+1}) \in HP(0, -\overrightarrow{W_{j_*}})$$

$$(z_1 + V_{i_*+1}) - W_{j_*+1} \in HP(0, \overrightarrow{W_{j_*}})$$

$$(z_1 + V_{i_*+1}) + (W_{j_*} - W_{j_*+1}) \in HP(W_{j_*}, W_{j_*} + \overrightarrow{W_{j_*}})$$

$$z_1 + V_{i_*+1} \in HP(W_{j_*}, W_{j_*+1})$$

which completes step (6).

Now we wish to show that $z_1 + V \subseteq HP(W_{j_*}, W_{j_*+1})$. (<u>Step (7)</u>:) Let $z_1 + V_{i_\#}$ be any vertex of $z_1 + V$, we will show that

$$z_1 + V_{i_\#} \in HP(W_{j_*}, W_{j_*+1}).$$

Either $\left\{\vec{V}_i\right\}_{i=i_*+1}^{i_\#-1}$ or $\left\{-\vec{V}_j\right\}_{i=i_\#}^{i_*} \subseteq HP(0,\vec{W_{j_*}})$. In the former case,

$$z_1 + V_{i_\#} = \left(\sum_{i=i_*+1}^{i_\#-1} \vec{V}_i\right) + z_1 + V_{i_*+1}, \text{ so:}$$

$$V_{i_\#} - V_{i_*+1} \in HP(0,\vec{W_{j_*}}).$$

But

$$z_1 + V_{i_*+1} \in HP(W_{j_*}, W_{j_*+1}),$$

so:

$$z_1 + V_{i_*+1} - W_{j_*} \in HP(0,\vec{W_{j_*}}),$$

$$z_1 + V_{i_\#} - W_{j_*} \in HP(0,\vec{W_{j_*}}) \text{ and}$$

$$z_1 + V_{i_\#} \in HP(W_{j_*}, W_{j_*+1})$$

which completes step (7).

Since we have shown that $z_1 + V \subseteq HP(W_{j_*}, W_{j_*+1})$ for arbitrary j_*, we have shown that

$$z_1 + V \subseteq \bigcap_{j=1}^{t} HP(W_j, W_{j+1}) = \text{con}(W).$$

Before we did step (6) we had shown that with $z_{j_0} = z_1$ we obtained $W_{j_0} \in (z_{j_0} + V)$. Therefore we are done. □

Notation: $_{z_0}K(z) \equiv K(z - z_0)$ for fixed $z_0 \in R^2$,

Proof of Theorem 2

If g is not identically zero pick a square $_nI^2_{i_*j_*}$ in the support of g such that W_{j_0} is a vertex of both $_nI^2_{i_*j_*}$ and W = the boundary of $con(supp(g))$. Pick z_1 according to Lemma 4. Then W_{j_0} is a vertex of the boundary of the $con(supp(_{z_1}K))$ and $_nI^2_{i_*j_*}$ is part of the support of $_{z_1}K$. Choose $b_1 = \pm g(z)$ for $z \in {_nI^2_{i_*j_*}}$ such that $g_1 \equiv g - b_1 {_{z_1}K}$ is zero on $_nI^2_{i_*j_*}$. Then g_1 has all the properties of g

(a) $g_1 \in Z(n)$,

(b) g_1 is a ghost with respect to $\{\theta_j\}_{j=1}^m$, and

(c) $con(supp(g_1)) \subseteq con(supp(g)) = con(W)$, but it also satisfies

(d) $con(supp(g)) \nsubseteq con(supp(g_1))$.

We can therefore pick another square but this time in the support of g_1, as we did for g. Apply Lemma 4 and obtain

$$g_2 \equiv g_1 - b_2 {_{z_2}K} \quad \text{such that}$$

$con(supp(g_2))$ is strictly contained in $con(supp(g_1))$.

This process can be repeated $L-1$ times (for some positive integer L) until the boundary of the $con(supp(g_{L-1}))$ = $z_L + V$, for some $z_L \in R^2$. (Lemma 3 says that this must happen.) Then Lemma 3 implies that there exists a $b_L \in R$ such that

$$g_L \equiv g_{L-1} - b_L \, z_L^K \quad \text{is identically zero.}$$

Thus

$$g = \sum_{i=1}^{L} b_i \, z_i^K,$$

The last conclusion of Theorem 2 is a consequence of Lemma 3.

□

CHAPTER VII

DEALING EFFECTIVELY WITH NOISY DATA

This chapter attacks the following issue: how should projection data be collected so that noise in the data will have the smallest effect on the choice of the reconstruction? In an experimental procedure, the data collected never exactly corresponds to

$$P_\theta f(z) = \int_{-\infty}^{+\infty} f(z + t\vec{\theta})dt , \text{ for } z \text{ in } \theta^\perp ;$$

the reasons for inaccuracies (i.e., noisy data) are as varied as there are experimental techniques. In any case, reconstruction in the presence of noise must be carefully handled because the knowledge that f is uniquely determined by $P_\theta f$ gives no indication about the effect on h caused by slightly inaccurate projection data. (Remember that we have assumed that the objective function, f, is in $Z(n)$ and that we are looking for a reconstruction of f, namely h, which is also in $Z(n)$.) Since the data is (almost) always inconsistent, (i.e, there may not exist a solution) some type of "averaging" method must be used to choose h. This chapter attacks the problem of choosing h with noisy data by using statistical methods involving the *Gauss-Markov theory* or the method of *Least Squares*.

Physical Justification of Importance and Sources of Noise

The reason that projection data is inaccurate is intimately connected with the original argument which establishes that the experimental problem can be modeled as a "reconstruction from projections" problem. For example: When an x-ray photon travels through a human body other things besides absorption may happen. The photon can be scattered by an atom; the photon can excite an electron which in turn emits another photon in an arbitrary direction; as well as other possibilities. As a result, the intensity of radiation detected at a point involves the mass distribution at points in the human body which can be far removed from the intended path of the radiation. When a radiograph is treated as a *projection* of the mass distribution, all of the aforementioned effects are considered to be factors which degrade the quality of the observations.

The Effect of Noisy Data on the Uniqueness of a Reconstruction

Any practical utilization of reconstruction from projections must recognize the difficulty posed by inaccurate (i.e., noisy) data. The following example demonstrates the effect of having noisy data on trying to obtain a good reconstruction.

On page 94 there is a **useful** diagram of a sequence of functions in $Z(24)$. The square with vertices O, A, B, C, represents $I^2 \subseteq R^2$. Since this diagram is only a contour map with the heights unlabeled, it requires further explanation: Let

solid	$g_1 \in Z(1)$	$g_1 = 1$ on $_1I_{11}^2$,	
dashed	$g_2 \in Z(2)$	$g_2 = 1$ on $_2I_{11}^2$	and $g_2 = -1$ on $_2I_{22}^2$,
dotted	$g_3 \in Z(3)$	$g_3 = 1$ on $_3I_{11}^2$	and $g_3 = -1$ on $_3I_{32}^2$,
solid	$g_4 \in Z(4)$	$g_4 = 1$ on $_4I_{11}^2$	and $g_4 = -1$ on $_4I_{43}^2$,
dashed	$g_6 \in Z(6)$	$g_6 = 1$ on $_6I_{11}^2$	and $g_6 = -1$ on $_6I_{65}^2$,
dotted	$g_8 \in Z(8)$	$g_8 = 1$ on $_8I_{11}^2$	and $g_8 = -1$ on $_8I_{87}^2$,
solid	$g_{12} \in Z(12)$	$g_{12} = 1$ on $_{12}I_{1,1}^2$	and $g_{12} = -1$ on $_{12}I_{12,11}^2$,
dashed	$g_{24} \in Z(24)$	$g_{24} = 1$ on $_{24}I_{1,1}$	and $g_{24} = -1$ on $_{24}I_{24,23}^2$.

All of the g_i's are zero where unspecified.

Also drawn on this diagram is a **solid** line which is inclined at the angle $\theta = \arctan(24/25)$ with respect to the x-axis. The significance of this line is that we are now able to consider $P_\theta(g_i)$ for $i = 2,3,4,6,8,12$, and 24. Because all of the g_i's are in $Z(24)$, we can apply Theorem 2 to realize that $P_\theta\big|_{Z(24)}$ has a trivial nullspace. As a measure of stability, we can hope to establish:

$$\|g_i\|_{L^2(I^2)} \leq M_2 \, \|P_\theta(g_i)\|_{L^2(\theta^\perp)} \; , \; \text{for some constant } M_2.$$

(Here, $\|k\|_{L^2(D)} = \left(\int_D |k(z)|^2 \, dz\right)^{1/2}$, where $k: D \to R$ and $D \subseteq R^p$, i.e., D is a subset of the p dimensional Euclidean space.)

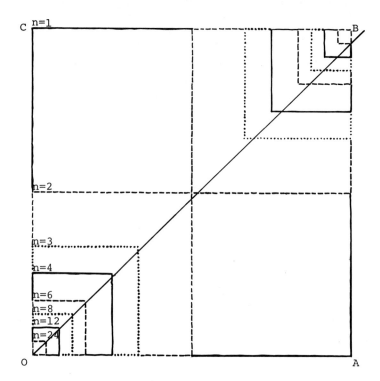

One can verify that the following values are approximately correct:*

g_i	$\|g_i\|_{L^\infty(I^2)}$	$\|P_\theta g_i\|_{L^\infty(\mathbf{R}^2)}$	$\|g_i\|_{L^2(I^2)}$	$\|P_\theta g_i\|_{L^2(\mathbf{R}^2)}$
g_2	1	.0297	.707	.029
g_3	1	.44	.467	.44
g_4	1	.316	.350	.31
g_6	1	.1934	.233	.193
g_8	1	.1282	.175	.128
g_{12}	1	.0952	.116	.095
g_{24}	1	.00504	.058	.005

Notice that for many of the g_i's, M_2 is not required to be very large, but g_{24} requires that M_2 be greater than 11 and g_2 requires that M_2 be greater than 23. We can see that very bad instability can occur in a reconstruction technique using $m = 1$, $\theta = \arctan(24/25)$ and $Z(n) = Z(24)$, since the addition of g_2 to a possible reconstruction has little effect on the projection data.

---

*$\|k\|_{L^\infty(D)} = \max_{z \in D} |k(z)|$, where $k: D \to R$ for $D \subseteq R^p$.

This example has demonstrated that careless application of Theorem 2 can be very dangerous in the presence of noisy data. In this case $\{\theta_j\}_{j=1}^{m} = \left\{\arctan \frac{24}{25}\right\}$ is the set of projection angles and $Z(24)$ is the reconstruction space. Theorem 2 says that reconstruction is unique in this case. (Since $n = 24 < 26 = 1 + \max(24, 25) = 1 + \max\left(\sum_{j=1}^{m} |p_j|, \sum_{j=1}^{m} |q_j|\right)$.) However, when there is noise in the projection data as insignificant as .029, measured in the L^2 topology, the resulting reconstruction can be off by the addition of $g_2 = X_{2^{I_{11}^2}} - X_{2^{I_{22}^2}}$. That is, if h is any proposed reconstruction of f such that

$$\|P_\theta f - P_\theta h\|_{L^2} \geq .029$$

then $h \pm g_2$ may be an equally good reconstruction of f.

The Effect of Noise on the Consistency of the Data

In general the number of individual readings of the projection data far exceeds n^2, the number of unknowns needed to specify h. Such a system is often *over determined* since the resulting number of equations is greater than the number of unknowns. Therefore, when the data is inaccurate, it is possible (in fact, probable) that no element of $Z(n)$ has the required projection data, i.e., to perfectly fit the observations. This situation is analogous to trying to find a line by knowing three points through which it must pass. If there is no possibility of incorrect data then any two points are quite sufficient, but if there is likely to be some error in the position of one of the points then no particular line is clearly preferable.

The Use of Least Squares - Advantages and Difficulties

The particular problem we must solve can be expressed using the matrix representation given in Chapter IV. Each projection, $P_{\theta_j} f$, is described by $U_{\theta_j} A = W_{\theta_j}$. If we want $P_{\{\theta\}} f$ to determine f then the uniqueness result (i.e., Theorem 2) tells us what system,

<u>VII.1</u>
$$\left\{ U_{\theta_i} A = W_{\theta_i} \right\}_{i=1}^{m},$$

determines A. (A is identified with f.) If this is done then

$$U = \begin{pmatrix} U_{\theta_1} \\ U_{\theta_2} \\ \cdot \\ \cdot \\ \cdot \\ U_{\theta_m} \end{pmatrix}$$

has rank equal to the number of its columns, i.e., $\text{rank}(U) = n^2$. Since the equation $U A = W$ is usually *inconsistent* (because

W is observed), the problem must be reformulated. The technique called Least Squares suggests solving:

$$\boxed{U^T U A = U^T W}$$

The choice of this method can be justified in light of facts like

(1) it seems to be the first choice of numerical analysts, and
(2) it may be the simplest way of handling inconsistent data.

This last equation is called the normal equation of the system $U A = W$. Such a reformulation of the problem choses the unique A which minimizes:

$$\| U A - W \|_{L^2} \quad .$$

(That is, the two vectors, $U A$ and W, are compared componentwise so that $\left(\sum_i | (U A)_i - (W)_i |^2 \right)^{1/2}$ is minimized.)

The obvious advantages of transforming the problem to solving the normal equation, $U^T U A = U^T W$, is that: (1) the new system is no longer inconsistent, and (2) a unique solution exists when U has full rank. These are clearly desirable effects, but now there is a new problem.

In the absence of error, the system VII.1 has a solution independent of any linear transformation of the data and the observational equations. However, the solution to the normal equation does not have this property. That is to say, performing Least Squares on a linear function of the data may not lead to the same answer. This is roughly because weighting different observations unequally can change the answer.

For example

Suppose we are looking for the line $ax + b = y$ in $R \times R$ which passes through the three points (x_1, y_1), (x_2, y_2) and (x_3, y_3). It is clear that a linear system which expresses the observational equations will be inconsistent in general (unless the three points are collinear). Consider two sets of observational equations:

$$\begin{Bmatrix} a x_1 + b = y_1 \\ a x_2 + b = y_2 \\ a x_3 + b = y_3 \end{Bmatrix} \quad \text{and} \quad \begin{Bmatrix} c a x_1 + c b = c y_1 \\ a x_2 + b = y_2 \\ a x_3 + b = y_3 \end{Bmatrix} ;$$

they can be represented by:

$$Q_1 A = Y_1 \quad \text{and} \quad Q_2 A = Y_2 \quad \text{where}$$

$$Q_1 = \begin{pmatrix} x_1 & 1 \\ x_2 & 1 \\ x_3 & 1 \end{pmatrix}, \quad Q_2 = \begin{pmatrix} cx_1 & c \\ x_2 & 1 \\ x_3 & 1 \end{pmatrix}, \quad A = \begin{pmatrix} a \\ b \end{pmatrix},$$

$$Y_1 = \begin{pmatrix} y_1 \\ y_2 \\ y_3 \end{pmatrix}, \quad \text{and} \quad Y_2 = \begin{pmatrix} cy_1 \\ y_2 \\ y_3 \end{pmatrix}.$$

Notice that the equations

$$Q_1^T Q_1 A = Q_1^T Y_1 \quad \text{and} \quad Q_2^T Q_2 A = Q_2^T Y_2$$

will have different solutions when $c \neq 1$. The datum (x_1, y_1) will be given *less weight* in this comparison if $|c| < 1$. As $|c|$ is reduced to zero, the effect of (x_1, y_1) on the choice of (a,b) will be decreased to zero accordingly.

In conclusion, the Least Squares technique provides a method for choosing a unique reconstruction even when there is noisy data (as long as U has rank $= n^2$). However, the particular reconstruction determined by $U^T U A = U^T W$ depends on arbitrary weighting factors within the original matrix equation, $U A = W$.

Statistical Considerations Relevant to the Use of Least Squares[*]

For experimental purposes we should consider the possibility of weighting our observations so that the data with the least error are given the greatest effect on the choice of the unknown vector A (= the reconstruction). These considerations have lead to the idea of finding a systematic way of assigning weights which depend in an appropriate way on the choice of the projection angles.

This question has lead the author to the Gauss-Markov

---

[*] I am indebted to Professor Charles Antoniak, Department of Statistics, Berkeley and Professor Henry Wynn, Department of Statistics, Imperial College, London for guidance in the choice of both the general as well as the particular statistical methods employed.

theory. A good reference is Rao [64]. (For readers not familiar with the necessary language of statistics, a brief description of relevant definitions and results is provided in an appendix to this chapter.)

If W are our observations, then our *statistical assumption* is that the coordinates of W are a set of random variables for which:

VII.2
(i) $E(W) = U A$ and
(ii) $Cov(W) \equiv D(W) \equiv \sigma^2 G$,

for some fixed constant σ **and matrix** G.

Then, the weighted least squares method (Gauss-Markov) leads to

VII.3
$$\boxed{U^T G^{-1} U A = U^T G^{-1} W}$$.

Any solution to VII.3 minimizes the quadratic form:

$$(W - UA)^T G^{-1} (W - UA) .$$

We call such a solution $\hat{A}$; it will exist and be unique provided that U and G have full rank. The quantity $\hat{A}$ is an *estimate of the unknown reconstruction*, and under the assumptions above, its dispersion is a measure of its *stability*. The dispersion matrix of $\hat{A}$ is given by:

$$\boxed{D(\hat{A}) = \sigma^2 (U^T G^{-1} U)^{-1}}$$.

In conclusion, the Gauss-Markoff method provides for an algorithm for choosing the best Least Squares equation. But in order to do this, one must be able to 1) determine $\sigma^2 G = D(W)$, the dispersion matrix of the projection data, and 2) evaluate the resulting dispersion matrix of the estimate of the unknown reconstruction.

Optimizing the Stability of the Estimate of the Unknown Reconstruction

Given that we have control over the U , (since U as described in Chapter IV can be manipulated by varying the projection angles) it is natural to ask whether there are better choices of U in some sense. This is exactly the statistical problem of experimental design. Recent work in the theory of optimum designs makes use of the criterion defined by a *loss function:*[*]

$$\boxed{\text{loss function} \equiv \det (U^T G^{-1} U)^{-1}}$$

The idea is that when the loss function is small, so is $(U^T G^{-1} U)^{-1}$ (in some sense), and we can infer that $\hat{A}$ has a small dispersion matrix. (The justification of this choice of the definition of the loss function includes: (1) this is a "natural generalization of uninvariant variance" and (2) this choice is invariant under linear transformations of the model.) Therefore, *by keeping the loss function small we can insure the stability of the estimate of the unknown*

---
[*]This loss function was introduced by Wald, (1943). [79]

reconstruction.

Choosing the Best Projection Angles

We can now present a set of reasonable assumptions which will allow for the choice of the best set of projection angles so as to maximize the stability of the estimate of the unknown reconstruction.

(1) Assume that the errors in the projection data as sampled are independent. (We mean that the errors are independent both between projections and within projections. That is, if ε_{θ_j} is the vector whose i^{th} component is the error in the ith measurement of the jth projection, then

(a) $\{\varepsilon_{\theta_1}, \varepsilon_{\theta_2}, \ldots, \varepsilon_{\theta_m}\}$ is statistically independent as a set of random vectors and

(b) the set of components of ε_{θ_j}, $\{(\varepsilon_{\theta_j})_1, (\varepsilon_{\theta_j})_2, \ldots, (\varepsilon_{\theta_j})_k\}$, is statistically independent as a set of random variables.)

(2) Assume that the variance in each projection is σ^2. That is, the variance of the error in a single measurement on a resolution element will vary inversely with the size of that element.

(3) Since every observation is interpreted to be

$$\sum q_{ij} a_{ij} + \text{error}$$

where the a_{ij}'s are coordinates of A and the q_{ij}'s are some positive numbers, we can normalize the calculations by assuming that the variance of this error is $\sum q_{ij}$. (This is justified because in this model we expect that the variance in the errors in the data resulting from a line of interactions is proportional to the total mass encountered along that line; in this case, we have uniformly normalized the density.)

If we let ε be the vector of errors in the observations then $W = UA + \varepsilon$. The first assumption says that $D(\varepsilon)$ is a diagonal matrix; since VII.2 implies that $E(\varepsilon) = 0$ and $D(W) = D(\varepsilon)$, we can infer that $D(W)$ is also diagonal. The second assumption says that the appropriate weighting must be that $U_{\theta_i} A = W_{\theta_i}$ and $U_{\theta_j} A = W_{\theta_j}$ are given the same weighted effect on the choice of A for any $i, j \leq m$. Therefore, $D(W_{\theta_j})$ only depends on the equation $U_{\theta_j} A = W_{\theta_j}$ and of course $D(W_{\theta_j})$ is diagonal. Finally, the last assumption says that the variance of the ith entry of W_{θ_j} is just:

$$\boxed{\sigma^2 \cdot \frac{1}{n^2} \cdot \sum_{k=1}^{n^2} (U_{\theta_j})_{ik}}$$

Now since σ^2 is just a constant of proportionality, it can be assumed to be 1 so that:

$$G = D(W) = \begin{bmatrix} D(W_{\theta_1}) & & & & 0 \\ & D(W_{\theta_2}) & & & \\ & & \cdot & & \\ & & & \cdot & \\ & & & & \cdot \\ 0 & & & & D(W_{\theta_m}) \end{bmatrix}$$

where $D(W_{\theta_j})$ is itself a diagonal matrix with $(i,i)^{th}$ entry equal to the *variance of the ith entry of* W_{θ_j}, i.e.,

$$v((W_{\theta_j})_i) = \frac{1}{n^2} \sum_{k=1}^{n^2} (U_{\theta_j})_{ik}.$$

Since $U_\theta A = W_\theta$ has been defined for all $\vec{\theta} \in S^1$ in Chapter IV, a computer program can be written which will minimize $\det(U^T G^{-1} U)^{-1}$ over any set containing values of $(n, m, \theta_1, \theta_2, \ldots, \theta_m)$.

For example:

Let $n = 2$, $m = 1$ and $h \in Z(n)$ given by

$$h = \sum_{i,j=1}^{2} a_{ij} X_{nI_{ij}^2}.$$ Suppose that $\theta = \arctan \frac{1}{2}$, then:

$$U = U_\theta = \begin{pmatrix} \frac{1}{4} & 0 & 0 & 0 \\ \frac{1}{2} & \frac{1}{4} & 0 & 0 \\ \frac{1}{4} & \frac{1}{2} & \frac{1}{4} & 0 \\ 0 & \frac{1}{4} & \frac{1}{2} & \frac{1}{4} \\ 0 & 0 & \frac{1}{4} & \frac{1}{2} \\ 0 & 0 & 0 & \frac{1}{4} \end{pmatrix}, \quad W_\theta = \begin{pmatrix} (W_\theta)_1 \\ (W_\theta)_2 \\ (W_\theta)_3 \\ (W_\theta)_4 \\ (W_\theta)_5 \\ (W_\theta)_6 \end{pmatrix} \quad \text{and}$$

$$A = \begin{pmatrix} a_{12} \\ a_{11} \\ a_{22} \\ a_{21} \end{pmatrix} \quad \text{so} \quad U_\theta A = W_\theta.$$

Then using the assumptions we get:

$$D(W) \equiv G \equiv \begin{pmatrix} 1/16 & 0 & 0 & 0 & 0 & 0 \\ 0 & 3/16 & 0 & 0 & 0 & 0 \\ 0 & 0 & 1/4 & 0 & 0 & 0 \\ 0 & 0 & 0 & 1/4 & 0 & 0 \\ 0 & 0 & 0 & 0 & 3/16 & 0 \\ 0 & 0 & 0 & 0 & 0 & 1/16 \end{pmatrix}.$$

After some calculations which can be simplified by exploiting the symmetry of the matrix we get:

$$\det(U_\theta^T G^{-1} U_\theta)^{-1} = 9/28 \approx .321 .$$

Therefore, for our defined loss function, $\tilde{F}(\theta) = \det(U^T G^{-1} U)^{-1}$, we get one evaluation: (when n = 2 and m = 1)

$$\tilde{F}(\arctan \tfrac{1}{2}) = 9/28 \approx .321 .$$

After much work the loss function has been evaluated at three other angles, (when n = 2 and m = 1):

$\tan\theta$	$\tilde{F}(\theta) = \det(U_\theta^T G^{-1} U_\theta)^{-1}$		θ
$\tfrac{1}{5}$	$\tfrac{1125}{1568}$	$\approx .719$	11.5°
$\tfrac{1}{4}$	$\tfrac{18}{35}$	$\approx .51$	13.0°
$\tfrac{1}{3}$	$\tfrac{81}{224}$	$\approx .362$	18.4°
$\tfrac{1}{2}$	$\tfrac{9}{28}$	$\approx .321$	28.6°

When this data is plotted in a graph of $\tilde{F}: (0°, 45°) \to R$ we can guess that a local minimum occurs around the midpoint of the domain. Through a rough argument, "interpolation" of the data can be done to find the local minimum. The result was that

$$\tilde{F}(\arctan 4/9) = \frac{2187}{7168} \approx .306 ,$$

which looks right and could be the local minimum. (Arctan 4/9 ≈ 24°.)

To summarize this example, suppose that we have decided to reconstruct an objective function by finding the best reconstruction in Z(2), and that we will only use one projection angle to obtain the data, then the projection angle which provides the optimally stable (with respect to our loss function) reconstruction technique seems to be arctan 4/9.

Conclusion to Chapter VII

Because of noise in the data, only inconsistent and inaccurate projection information is available. A least squares approach is used to overcome this difficulty; the Gauss-Markov method, which is a statistical treatment of least squares, is explained. Finally, the problem of maximizing the stability of an estimate of the unknown reconstruction leads naturally to an optimum experimental design problem, which can be utilized to chose the "best" projection angles.

APPENDIX TO CHAPTER VII

STATISTICAL REFERENCE MATERIAL

The basis of statistics is probability theory which has many features in common with measure theory of mathematics. The fundamental concept of statistics is the random variable: once its definition is clear, other ideas can be explained easily.

Definition: A *real valued random variable X(·)* is just a real valued measurable function on the measure space (Ω, B, P) where $X(\cdot)$ is defined on $\omega \in \Omega$, B is the Borel sets on Ω, and P is a positive measure on Ω such that $P(\Omega) = 1$.

Definition: The *distribution function F(x) of a random variable X* is a function on R such that

$$F(x) = P(\{\omega \in \Omega : X(\omega) \leq x\}) = P(X^{-1}(-\infty, x)).$$

We will assume that $F(x)$ is absolutely continuous with respect to Lebesgue measure so that the Radon-Nikodym Theorem can be used to get the *probability density function f(x) for the random variable X(·)* which is defined by

$$F(x) = \int_{-\infty}^{x} f(t)dt \quad \text{where} \quad dt \text{ is Lebesgue measure.}$$

Thus when we have

$$P(X^{-1}(-\infty,x)) = \int_{-\infty}^{x} f(t)dt ,$$

we get the mathematical statement:

$$\left\{\begin{array}{l} \text{There exists } X(\cdot), \text{ a real} \\ \text{valued random variable} \\ \text{defined on the probability} \\ \text{space } \Omega . \end{array}\right\} \iff \left\{\begin{array}{l} \text{There exists a positive} \\ \text{function, } f \in L^1(R) \\ \text{such that} \\ \|f\|_{L^1} = 1 . \end{array}\right.$$

Let X_1 and X_2 be real valued random variables, then $aX_1 + X_2$, and $aX_1 X_2$ are random variables for $a \in R$. In practice, it is often necessary to consider pairs of random variables (X_1, X_2), i.e., a vector whose components are random variables -- this is called a *two dimensional random variable*. It satisfies all the previous definitions except everywhere R was written replace it by R^2. Then the probability density function for (X_1, X_2) is of the form $f(t_1, t_2)$, i.e., *f is the probability density function for* (X_1, X_2) iff $f \in L^1(R^2)$, $\|f\|_{L^1(R^2)} = 1$ and

$$\int_{-\infty}^{x_1} \int_{-\infty}^{x_2} f(t_1,t_2)dt_2 dt_1 = P(X_1^{-1}(-\infty,x_1) \cap X_2^{-1}(-\infty,x_2)).$$

The average value of a random variable X is the same as the *expectation value of a random variable X* and is denoted by

$$E(X) \equiv \int_{-\infty}^{+\infty} tf(t)dt = \|xf(x)\|_{L^1(R)}$$

where $f(x)$ is the probability density for X. This is why $E(X)$ is sometimes called the *first moment of* X, but it is actually the first moment of the probability density function for X. The notation $\mu_x \equiv$ mean of $X \equiv E(X)$ is often used.

Variance of a random variable is the second moment about the mean of the probability density function. $\sigma^2 \equiv$ variance $\equiv v(X)$.

$$\sigma^2 = \int_{-\infty}^{+\infty} (t - \mu_x)^2 f(t)dt = \|(x - \mu_x)^2 f(x)\|_{L^1(R)}$$

Two random variables can be "compared" via the device called the *covariance* $\equiv cov(X_1, X_2)$. Here the probability density function for (X_1, X_2) is used:

$$cov(X_1, X_2) \equiv \int_{-\infty}^{+\infty} \int_{-\infty}^{+\infty} (t_1 - \mu_1)(t_2 - \mu_2) f(t_1, t_2) dt_1 dt_2 \ .$$

Statisticians have shown that a random variable can be well estimated from just knowing its mean and variance. Therefore, it is upon these quantities that much statistical work is done.

Two random variables are called *independent* when $f(x_1, x_2) = f(x_1)f(x_2)$. By the formula for $\text{cov}(X_1, X_2)$ it is clear that if X_1 and X_2 are independent then $\text{cov}(X_1, X_2) = 0$.

The *dispersion matrix* or *covariance matrix* of an n dimensional vector $(X_1, X_2, \ldots, X_n) = \underline{X}$ is

$$D(\underline{X}) \equiv \text{cov}(\underline{X}) \equiv (\text{cov}(X_i, X_j)), \quad \text{i.e.,}$$

$D(\underline{X})$ is an $n \times n$ matrix with i,j^{th} entry $\text{cov}(X_i, X_j)$. When $X_1, X_2, \ldots, X_n$ are independent random variables then $D(\underline{X})$ is a diagonal matrix and $D_{ii}(\underline{X})$ = variance of X_i.

In general, we can consider the inner product space over the reals with basis $\{(0,0,\ldots,0,X_i,0,\ldots,0)\}_{i=1}^{n}$ and inner product given by the symmetric positive definite matrix $D(\underline{X})$. Then we can easily calculate the variance and covariance of linear functionals of $\underline{X}$:

Let $L^T = (\ell_1, \ell_2, \ldots, \ell_n)$, and let $L^T \underline{X}$ mean $\sum_{i=j}^{n} \ell_i X_i$, similarly for $M^T = (m_1, m_2, \ldots, m_n)$.

$$v(L^T \underline{X}) \equiv L^T D(\underline{X}) L \equiv <L, L> \quad \text{and}$$

$$\text{cov}(L^T \underline{X}, M^T \underline{X}) \equiv L^T D(\underline{X}) M \equiv <L, M>.$$

Therefore, we get:

$$\text{cov}(L^T \underline{X}) = L^T \text{cov}(\underline{X}) L.$$

CHAPTER VIII

HOW A RECONSTRUCTION APPROXIMATES A REAL LIFE OBJECT

This chapter addresses the problem of reconstruction from projections in the context of reasonable assumptions about the original objective function. Since the objective function is never in the pixel reconstruction space, i.e., $Z(n)$, it is not sufficient to resolve the problem by simply showing (as was done in Chapters III through VII) that if f is in $Z(n)$, then the problem of reconstruction from projections can be made well-posed. Within this chapter, f is not assumed to be in $Z(n)$, and the following question is answered: *How close is* h *to* f *when* h *is chosen according to the criterion that* $P_{\{\theta\}}h$ *is closest to* $P_{\{\theta\}}f$, *among all possible* h *in* $Z(n)$?

First, *assumptions* or choices are made concerning reasonable a priori information about f and the appropriate definition of closeness (i.e., topology of the function space containing f) and etc. Second, the assumptions are shown to have five important *consequences*. Then, these consequences are used to *estimate* h - f in the chosen topology, in this way presenting a theory of reconstruction from projections.[*] Finally, this estimate is shown to have much practical *significance* and many physical *applications*.

---

[*]The first two pages of Chapter X present a general formulation of an estimate for h - f. The calculation of Chapter VIII is just a special case of that formulation.

Assumptions with their Justifications

This section presents the fundamental assumptions which allow the following important goal to be reached: *estimating the closeness between the objective function,* f, *and the reconstruction,* h. Basically, this amounts to the choice of three function spaces:
(1) the reconstruction space,
(2) the infinite dimensional space containing f, and
(3) the infinite dimensional space containing the projection data.

Assumption 1: The reconstruction space is $Z(n)$.

This choice is made for two reasons. It is the most commonly used reconstruction space, and this is the space about which we have all the necessary information, that is, we know how to make the problem of reconstruction from projections *well-posed*[*] when we use this reconstruction space.

Assumption 2: The following three assumptions concern f.

---

[*]See footnote on last page of Chapter II.

2a: $f = F * g_0$ where F represents the actual distribution of mass at the resolution of atomic distances[*] and g_0 is a C^∞ bump function.[**]

The fact that we are trying to solve a real problem provides us with usable *a priori* information. Because we are sampling a real object with an X ray beam, perfect data is not actually like

$$P_\theta F(z) = \int_{-\infty}^{+\infty} F(z + t\vec{\theta}) \, dt \; ;$$

an X ray photon is not only affected by the matter encountered along a line $\{z_0 + t\vec{\theta} : t \in R\}$, but is also affected by matter at small distances from that line. A probablistic treatment can be made to justify that the function F which actually represents the coefficient of X ray attenuation within a head is not measured by its X ray projection. In fact,

<u>VIII.1</u> $f(z) = (F * g_0)(z) \equiv \int_{-\infty}^{+\infty} F(z') g_0(z - z') dz'$

is the function measured by an X ray projection of F, where g_0 is a C^∞ bump function concentrated at the origin. Engineers might choose $g_0(z) = (\sin |z|\sigma^{-2})/(|z|\sigma^{-2})$,

---

[*] That is, F contains information at a resolution of about one Angstrom.

[**] By a C^∞ bump function, we mean a real valued function defined on R^2 which is infinitely differentiable, close to zero outside of some compact subset $\Omega \subseteq R^2$, and close to unity on Ω, i.e., a smooth function which is concentrated at Ω.

while it could be argued that a gaussian, $g_0(z) = e^{-2\pi z^2 \sigma^{-2}}$, is more appropriate.

2b: Some approximation to the following function is known: $A(t) = \text{area}(\{z \in I^2 : \|\nabla F(z)\| \geq t\})$. ($A: R^+ \to [0,1]$.)

This information just says that some knowledge about the actual mass distribution is known. In particular, consider a cross-section of a head: at the skull there are great changes in the density, while in the brain matter much smaller changes are present. Knowing $A(t)$ in the case of a cross-section of a head means **knowing** what percentage (i.e., fraction of the area) of such a cross-section has the property that great local changes in density exist. This information is available in an *a priori* sense since a good approximation of $A(t)$ can be found which is true of all cross-sections of human brains.

2c: $f \in L^2(I^2)$.

This assumption says that $\left(\int_{I^2} |f(x,y)|^2 \, dxdy \right)^{1/2}$ is finite, but more significantly, it represents a choice of the definition of closeness which is used to compare h with f.

The choice of this function space has been carefully considered. It is necessary that $Z(n)$ be contained in the function space used because we want to eventually compare h to f . A space of continuous functions was considered but its major draw back is that Z(n) contains many functions which are not continuous. However, Z(n) is a space of piecewise continuous functions; therefore, we needed a function space which naturally contains such functions. Any of the L^p spaces could have been used since they contain the piecewise continuous functions and are complete.* L^2 was chosen because of its additional structure, i.e., inner-product.

Finally, the choice of the L^2 topology is an excellent way to compare h and f since

$$\|h - f\|_{L^2} = \left(\int_{I^2} |h(x,y) - f(x,y)|^2 \, dxdy \right)^{1/2} .$$

This kind of comparison is not terribly obscured by the fact that h is full of discontinuities. In fact, the integral does not even consider the value of h — f along the lines of discontinuity of h , i.e., the lines
$\{(x,j/n): j \in Z$ and $0 \le j \le n, x \in R\}$ and
$\{(i/n,y): i \in Z$ and $0 \le i \le n, y \in R\}$. An important quality of this topology is the fact that all differences between h and f contribute to $\|h - f\|_{L^2}$ in a positive way.

---

*A metric space is *complete* if every Cauchy sequence converges to an element of the space.

Assumption 3: The projection data,

$$P_{\{\theta\}}f = \left(P_{\theta_1}f, P_{\theta_2}f, \ldots, P_{\theta_m}f\right), \text{ is an element of}$$

$$L^2\left(\bigoplus_{j=1}^{m} \theta_j^{\perp}\right).$$

This choice of the topology of some space of functions defined on $\bigoplus_{j=1}^{m} \theta_j^{\perp}$ satisfies two valuable characteristics. First of all, it is compatible with the utilization of the Least Squares technique described in Chapter VII. Secondly, choosing this topology is consistent with the facts: 1) observed data are generally noisy and 2) naturally occuring noise tends to have a small L^2 norm.

Consequences of the Assumptions

Consequence 1: There is a unique $\bar{f}_n$ which is the closest element of $Z(n)$ to f, i.e., $\bar{f}_n \in Z(n)$ is determined by

VIII.2 $\qquad \|\bar{f}_n - f\|_{L^2} \leq \|k - f\|_{L^2} \quad \text{for all} \quad k \in Z(n)$.

The following proposition shows that $\bar{f}_n$ is well-defined and that it also has an important *average value* characteristic.

Proposition VIII.1 Let $f \in L^2(I^2)$, and let n be a particular positive integer, then there is a unique $\bar{f}_n \in Z(n)$ which satisfies VIII.2. Furthermore, if

$$\bar{f}_n = \sum_{i,j=1}^{m} b_{ij}\, \chi_{{}_nI^2_{ij}} \quad \text{then}$$

$$b_{ij} = n^2 \int_{{}_nI^2_{ij}} f(x,y)\, dxdy\,, \quad \text{i.e.,}$$

b_{ij} is the average value of f on the pixel ${}_nI^2_{ij}$.

Proposition VIII.1 says that the same element of $Z(n)$ which is defined by the natural *least squares* criterion for closeness to f, also happens to be that function which is the *average value* of f on each pixel. This average value quality of $\bar{f}_n$ gives further significance to $\bar{f}_n$ since, even if there were no link between least squares and average value, the particular element of $Z(n)$ which takes the average value of f on each pixel is also a good choice for the best reconstruction of f.

Proposition VIII.2 Let $T_n: L^2(I^2) \to L^2(I^2)$ be defined by $T_n f = \bar{f}_n$, then $T_n \to I$ in the strong operator topology of L^2.[*]

---
[*] I am indebted to William Bade, Professor of Mathematics, Berkeley, for the proof of this result.

Consequence 2: $\|f - \bar{f}_n\|_{L^2} \leq \varepsilon_1(n)$ can be estimated.

The second consequence points out that it is possible to predict how close $\bar{f}_n$ is to f. By using the *a priori* information about f which was presented in Assumption 2, we can provide the hypotheses for two propositions which explain this estimate.

Proposition VIII.3   If $f \in C^1(I^2)$ and ${}_nB_{ij} \geq \|\nabla f\|_{L^\infty}({}_nI_{ij}^2)$ then

VIII.3 $\|f - \bar{f}_n\|_{L^2}({}_nI_{ij}^2) \leq n^{-2} \cdot {}_nB_{ij}$. *†

Proposition VIII.4   If $f = F * g$ and $F \in C^1(I^2)$, then

VIII.4 $\|\nabla f\|_{L^\infty}({}_nI_{ij}^2) \leq$

$\min\left(\|\nabla g\|_{L^\infty(R^2)} \|F\|_{L^2}({}_nI_{ij}^2), \|\nabla F\|_{L^\infty}({}_nI_{ij}^2) \|g\|_{L^2}({}_nI_{ij}^2)\right)$.

---

* $f \in C^1(I^2)$ means that f is defined on I^2 and that f's first partial derivatives are continuous.

† I am indebted to Kennan Smith for making this result available.

The reader should see that Proposition VIII.4 provides the bounds on the value for $_nB_{ij}$ used in Proposition VIII.3. Therefore,

VIII.5 $\varepsilon_1(n) \equiv$

$$\left(\sum_{i,j=1}^{n} (_nB_{ij})^2 \right)^{1/2} n^{-2} \geq \| f - \bar{f}_n \|_{L^2}$$

will do.

Fortunately $\varepsilon_1(n)$ decreases rapidly as n increases, since as n increases a larger and larger percentage of the $_nI_{ij}^2$'s will not overlie boundaries between large variations of f. The squares which overlie boundaries are numbered on the order of n, while the squares which do not overlie boundaries are numbered on the order of n^2.

For example: Let f_o be the characteristic function on the unit disk = { $z \in R^2$: $||z|| \leq 1$ } , i.e., $f_o(z) = 1$ if $||z|| \leq 1$ and zero elsewhere. The figure on the left is a contour graph of f_o as an element of $L^2(I^2)$; it also

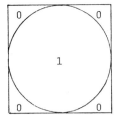

depicts an example when $n = 1$, i.e., $_1I^2_{11} = I^2$ overlies boundaries of f_o . Notice that when n increases, a higher and higher percentage of the $_nI^2_{ij}$'s do not overlie boundaries of variation of f_o . Let $E_1(n)$ = the number of squares, $_nI^2_{ij}$, which overlie boundaries of variation of f_o ; let $E_2(n)$ = the number of squares, $_nI^2_{ij}$, which do not overlie boundaries of variation of f_o .

n	$E_1(n)$	$E_2(n)$	Diagram
1	1	0	
2	4	0	
3	8	1	
4	12	4	
5	16	9	
6	20	16	
7	24	25	
.	.	.	
.	.	.	
.	.	.	
n	4(n - 1)	$(n - 2)^2$	

The advantage obtained by $E_2(n)$ dominating $E_1(n)$ is realized by returning to Proposition VIII.4. On squares which overlie boundaries of variation of F, $\|\nabla g_0\|_{L^\infty(R^2)}$, $\|F\|_{L^2(_nI_{ij}^2)}$ should be used to estimate $\|\nabla f\|_{L^\infty(_nI_{ij}^2)}$ while on squares which do not overlie boundaries of variation of F, $\|\nabla F\|_{L^\infty(_nI_{ij}^2)} \|g_0\|_{L^2(_nI_{ij}^2)}$ should be used. When we are concerned with reconstruction of cross-sections of heads, the second estimate may be very useful since a large majority of the pixels will overlie regions of brain matter which are characterized by small changes in density. This will result in smaller values for the $_nB_{ij}$'s and, therefore, a smaller value for $\varepsilon_1(n)$.

Consequence 3: $\|P_{\{\theta\}}f' - P_{\{\theta\}}f\|_{L^2} \leq M_1 \|f' - f\|_{L^2}$, for f and $f' \in L^2(I^2)$.

This is a restatement of the fact that the operator norm of $P_{\{\theta\}}$ exists and can be given by M_1. In fact, $M_1 = m^{1/2}$ since:

Proposition VIII.5 Let $\{\vec{\theta}_i\}_{i=1}^m \subseteq S^1$. Define $P_{\{\theta\}}: L^2(I^2) \to L^2\left(\bigoplus_{i=1}^m \theta_i^1\right)$ by $P_{\{\theta\}} = \left(P_{\theta_1}f, P_{\theta_2}f, \ldots, P_{\theta_m}f\right)$. Then $P_{\{\theta\}}$ is a continuous linear transformation with operator norm $= m^{1/2}$.*

---
*An equivalent result appears in [71].

Consequence 4:

$$\|h - k\|_{L^2} \leq M_2\left(n, \{\theta_j\}_{j=1}^m\right) \|P_{\{\theta\}}h - P_{\{\theta\}}k\|_{L^2},$$

for $h, k \in Z(n)$.

This results from the fact that $P_{\{\theta\}}: Z(n) \to L^2\left(\bigoplus_{i=1}^m \theta_i^\perp\right)$ can be represented by $UA = W$ according to Chapter IV. When Theorem 2 is applied so that U has rank $= n^2$, there exists a linear transformation $\bar{P}^{-1}: P_{\{\theta\}}(Z(n) \to Z(n)$ such that $\bar{P}^{-1}(P_{\{\theta\}}(k)) = k$ for all $k \in Z(n)$. It is well known that the operator norm of $\bar{P}^{-1}$ in L^2 can be expressed in terms of U; in fact, $M_2 = $ (smallest eigenvalue of $U^T U)^{-1/2}$. Since U is a function of n and $\{\theta_j\}_{j=1}^m$, so is M_2.

Consequence 5: $\varepsilon_0 \equiv \|P_{\{\theta\}}f - P_{\{\theta\}}h\|_{L^2}$ can be directly measured once the data, i.e., $P_{\{\theta\}}f$, is known and as soon as h is chosen.

This completes the list of the direct consequences of the assumptions.

<u>Estimating</u> $\|h - f\|_{L^2}$, <u>i.e.,</u>

<u>How close is the obtained reconstruction to the unknown objective function?</u>

The existence of $\bar{f}_n$ and the other four consequences allow for an estimate of $\|h - f\|_{L^2}$ since

$$\|h - f\|_{L^2(I^2)}$$

$$\leq \|f - \bar{f}_n\|_{L^2(I^2)} + \|\bar{f}_n - h\|_{L^2(I^2)}$$

$$\leq \varepsilon_1(n) + M_2 \|P_{\{\theta\}}\bar{f}_n - P_{\{\theta\}}h\|_{L^2\left(\bigoplus_{i=1}^{m} \theta_i^{\perp}\right)}$$

$$\leq \varepsilon_1(n) +$$

$$M_2 \left(\|P_{\{\theta\}}\bar{f}_n - P_{\{\theta\}}f\|_{L^2\left(\bigoplus_{i=1}^{m} \theta_i^{\perp}\right)} + \|P_{\{\theta\}}f - P_{\{\theta\}}h\|_{L^2\left(\bigoplus_{i=1}^{m} \theta_i^{\perp}\right)} \right)$$

$$\leq \varepsilon_1(n) + M_2 \left(M_1 \|f - \bar{f}_n\|_{L^2(I^2)} + \varepsilon_0 \right)$$

$$\leq \varepsilon_1(n) + M_2 \left(M_1 \varepsilon_1(n) + \varepsilon_0 \right).$$

Therefore, in our case:

<u>VIII.6</u> $\|f - h\|_{L^2} \leq \varepsilon_1(n) +$
(smallest eigenvalue of $U^T U)^{-\frac{1}{2}} \left(m^{\frac{1}{2}} \varepsilon_1(n) + \varepsilon_0 \right)$.

where U is given in Chapter IV for angles chosen according to Theorem 2.

For further clarification of the estimate consider the following diagram.

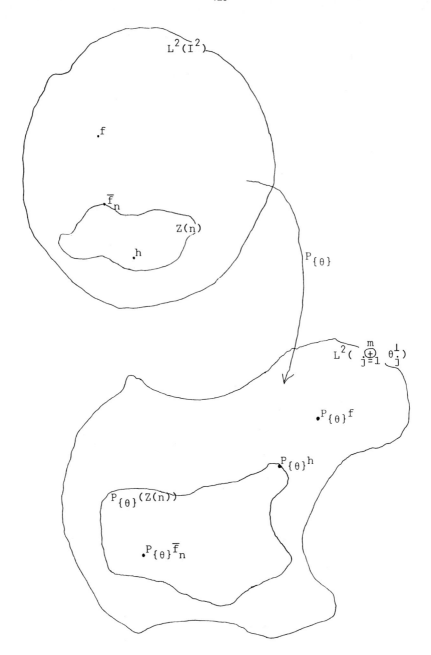

By formula <u>VIII.6</u> it is clear that $\|h - f\|_{L^2}$ can be estimated when $\varepsilon_1(n)$ is computed and ε_0 is measured. This fact agrees with K. Smith's result since $\varepsilon_1(n)$ represents the necessary *a priori* information without which no estimate of the reliability of a reconstruction can be made (no matter how small ε_0 is).

<u>Significance and Applications of the Estimate of</u> $\|h - f\|_{L^2}$

The estimate can be used to direct experimental and theoretical design. If one returns to the last line of the inequalities which estimate $\|h - f\|_{L^2}$, i.e.,

<u>VIII.7</u> $\quad \|h - f\|_{L^2} \leq \varepsilon_1(n) + M_2 (M_1 \varepsilon_1(n) + \varepsilon_0)$,

then a theoretical assumption can be made which allows for *a priori* estimation. Suppose that either ε_0 is proportional to $M_1 \varepsilon_1(n)$ or when compared to $M_1 \varepsilon_1(n)$ is so small as to be considered negligible. The supposition is justified since it then becomes possible to make some clear decisions about the theoretical choices of n, m and $\{\theta_j\}_{j=1}^m$. That is, the design of an experiment should be chosen in a way which is consistent with the desire to keep $\varepsilon_1(n)$, M_2 and m as small as possible (according to <u>VIII.7</u>).

Secondly, the inequality <u>VIII.7</u> provides an *a posteriori* estimate of the closeness between h and f. To the knowledge of the author, this is the first time that information of this type has been available. It is important to recognize that K. Smith's result predicts that no such estimate can be made, but of course, his result depended on an almost total lack of *a priori* information about f -- a situation which we have not assumed.

Perhaps the most significant effect of the fact that an a posteriori estimate can be made is that it is now possible to define the resolution of a reconstruction. As mentioned in Chapter III, researchers have differed about their definition of the resolution of a reconstruction. K. Smith recognizes that the size of the smallest component of h does not necessarily represent reliable information about f. However, it is not until the closeness between h and f is known that a clear and definitive answer to the resolution issue can be given.

Conclusions

The significance of this analysis extends both to the qualitative usefulness of any reconstruction from projections procedure and to the practical questions which must be confronted before any empirical work can be done.

In the past, reconstructions were empirically obtained on a fixed grid, say n × n (i.e., the reconstruction is an element of $Z(n)$). Many things about this reconstruction were not known, including:

(1) Does there exist a reconstruction which is uniquely determined by the fact that it agrees with a particular set of noiseless projection data?

(ii) How is the choice of a particular reconstruction affected by the presence of noisy data?

(iii) How does the choice of the projection angles affect (i) and (ii)?

(iv) How close is an obtained reconstruction to an unknown objective function?

And finally,

(v) At what resolution can details of the unknown objective function be inferred from its reconstruction?

The reader should agree that each of these questions is fundamental to those that follow it. For example: If (i) were not true then (ii) would not make sense; if there are classes (which contain more than one element) of equally correct reconstructions for each set of noiseless projection data then there could not be a single "best" reconstruction from noisy data. Similarly, the design question, (iii), could not be considered outside of the favorable answers to questions (i) and (ii).

After the reconstruction space was explained in Chapter III (p.41), questions (i), (ii), and (iii) were answered in the next four chapters. In fact, Theorem 2 (p.64) shows how the projection angles can be critically important to the answer to question (i). While in the seventh chapter, the Gauss-Markov Theory is utilized to show how the stability of reconstruction depends on the choice of the projection angles when the data are considered to be noisy (p.100).

Once this groundwork had been laid, it became possible to make the estimate, <u>VIII.6</u>, which is an answer to question (iv). To the knowledge of the author, this is the first time that the closeness between an obtained reconstruction and an unknown objective function has been theoretically estimated. Moreover, it is done with full cognizance of Kennan Smith's result (p.23).

Finally, the question of available resolution about the objective function can be put on a firm footing. If, for example, our answer to question (iv) was that $\|h - f\|_{L^2}$ was negligibly small, then we can infer that 1/n, which is the (picture) resolution of h, must be equivalent to the available resolution about the objective function, f. On the other hand, if $\|h - f\|_{L^2}$ can only be very poorly estimated, then only information at a resolution which is much worse than 1/n, is provided about the objective function.

The empirical consequences of this work are dramatic since we have now demonstrated that it is no longer necessary to depend on trial and error to obtain consistently reliable reconstructions. In Chapter VII it was shown how to make h optimally stable, i.e., to force the error in h to have small variance, while in this chapter, it was shown exactly which parameters should be manipulated in order to insure that h is close to f.

Proofs of Propositions stated in Chapter VIII

Proof of Proposition VIII.1 Let $h_n \in Z(n)$ satisfy VIII.2.

Since $\|f - h_n\|^2_{L^2(I^2)} = \sum_{i,j=1}^{n} \int_{{}_nI^2_{ij}} (f(z) - h_n(z))^2 \, dz$,

we are done if we can show that $h_n = \bar{f}_n$ on each ${}_nI^2_{ij}$.

Let $r_0 = h_n(z)$ for z in ${}_nI^2_{i_0 j_0}$, then r_0 must

minimize $\|f - r\|^2_{L^2({}_nI^2_{i_0 j_0})} = \langle f - r, f - r \rangle =$

$\|f\|^2_{L^2({}_nI^2_{ij})} - 2r \int_{{}_nI^2_{ij}} f(z) dz + r^2 \int_{{}_nI^2_{ij}} dz$, i.e., a quadratic

in r. It attains its minimum at the vertex of the parabola,

i.e., $r_0 = n^2 \int_{{}_nI^2_{ij}} f(z) \, dz$. $\square$

Therefore $\bar{f}_n$ which was defined with the *least squares* property also satisfies the *average value* property. It should be recognized that this result breaks down if f takes its values in the complex plane, which is significant to the Fourier techniques.

Proposition VIII.2 $T_n \to I$ in the strong operator topology of L^p, for $p \in [1, \infty)$.

Proof: By the Banach-Steinhaus Theorem, we need the fact that $T_n f \to f$ in L^p for f continuous and must prove that $T_n f$ is a bounded sequence in $L^p(I^2)$ for arbitrary f. To see that $T_n f \to f$ in L^p for f continuous we can use the argument that $\bar{f}_n \to f$ in the uniform topology and recall that the uniform topology is a stronger topology than L^p for $1 \le p \le \infty$. This shows that $T_n f \to I(f) \equiv f$ on the continuous functions; since the continuous functions are dense in $L^p(I^2)$, the sequence of operators is determined on all of L^p if it can be shown that it is pointwise bounded: Let f be an arbitrary element of $L^p(I^2) [\subset L^1(I^2)]$, then:

$$|T_n f|^p = \sum_{i,j=1}^n \left| n^2 \int_{{}_n I_{ij}^2} f \, dz \right|^p |\chi_{{}_n I_{ij}^2}|^p$$

$$\|T_n f\|_p = \left(\sum_{i,j=1}^n \left| n^2 \int_{{}_n I_{ij}^2} f \, dz \right|^p \int_{R^2} |\chi_{{}_n I_{ij}^2}|^p \, dz \right)^{1/p}$$

$$= \left(\sum_{i,j=1}^n n^{2p} \left| \int_{{}_n I_{ij}^2} f \, dz \right|^p n^{-2} \right)^{1/p}$$

$$\le \left(\sum_{i,j=1}^n n^{2(p-1)} \left(\int_{{}_n I_{ij}^2} |f| \, dz \right)^p \right)^{1/p} .$$

Now use Hölder's inequality and note that
$1/q + 1/p = 1 \Rightarrow p/q = p-1$,

$$\leq \left(\sum_{i,j=1}^{n} n^{2p/q} \left(\left(\int_{{}_nI_{ij}^2} |f|^p dz \right)^{1/p} \left(\int_{{}_nI_{ij}^2} |1|^q dz \right)^{1/q} \right)^p \right)^{1/p}$$

$$= \left(\sum_{i,j=1}^{n} n^{2p/q} \int_{{}_nI_{ij}^2} |f|^p \, dz \, (n^{-2})^{p/q} \right)^{1/p}$$

$$= \left(\sum_{i,j=1}^{n} \int_{{}_nI_{ij}^2} |f|^p \, dz \right)^{1/p}$$

$$= \|f\|_p .$$

Therefore $\|T_n\|_p = 1$, for all n, and we are done. □

Proposition VIII.3 If $f \in C^1(I^2)$ and ${}_nB_{ij} \geq \|\nabla f\|$ on ${}_nI_{ij}^2$ then

$$\|f - \bar{f}_n\|_{L^2({}_nI_{ij}^2)} \leq {}_nB_{ij} \, n^{-2} .$$

Proof: This result follows immediately from the two lemmas proven below.

Lemma 1

$$\int_{{}_nI_{ij}^2} \int_{{}_nI_{ij}^2} (f(z_1) - f(z_2))^2 \, dz_1 dz_2 = 2n^{-2} \|f - \bar{f}_n\|_{L^2({}_nI_{ij}^2)}^2 .$$

Lemma 2

$$\int_{_nI_{ij}^2}\int_{_nI_{ij}^2} (f(z_1)-f(z_2))^2 \, dz_1 dz_2 \leq 2 \, _nB_{ij}^2 \, n^{-6}.$$

Proof of Lemma 1:

$$\int_{_nI_{ij}^2}\int_{_nI_{ij}^2} (f(z_1) - f(z_2))^2 \, dz_1 \, dz_2$$

$$= \int_{_nI_{ij}^2}\int_{_nI_{ij}^2} ((f(z_1)-\bar{f}_n(z_1)) - (f(z_2)-\bar{f}_n(z_2)))^2 \, dz_1 \, dz_2$$

$$= 2 \int_{_nI_{ij}^2}\int_{_nI_{ij}^2} (f(z_1)-\bar{f}_n(z_1))^2 \, dz_1 \, dz_2$$

$$- 2 \int_{_nI_{ij}^2}\int_{_nI_{ij}^2} (f(z_1)-\bar{f}_n)(f(z_2) - \bar{f}_n) \, dz_1$$

$$= 2n^{-2} \int_{_nI_{ij}^2} (f(z_1) - \bar{f}_n(z_1))^2 \, dz_1 + 0$$

$$= 2n^{-2} \, \|f - \bar{f}_n\|_{L^2}^2 \left(_nI_{ij}^2\right) \qquad \square$$

Proof of Lemma 2: Let $\vec{\theta} = (z_1 - z_2)/\|z_1 - z_2\|$, then

$$\int_{_nI_{ij}^2} \int_{_nI_{ij}^2} (f(z_1) - f(z_2))^2 \, dz_1 \, dz_2$$

$$= \int_{_nI_{ij}^2} \int_{_nI_{ij}^2} \left(\int_0^{\|z_1-z_2\|} \nabla f(z_2 + t\vec{\theta}) \cdot \vec{\theta} \, dt \right)^2 dz_1 \, dz_2$$

by the fundamental theorem of calculus. By Cauchy-Schwartz:

$$\leq \int_{_nI_{ij}^2} \int_{_nI_{ij}^2} \left(\int_0^{\|z_1-z_2\|} |\nabla f(z_2 + t\vec{\theta})| \, |\vec{\theta}| dt \right)^2 dz_1 \, dz_2$$

$$\leq {}_nB_{ij}^2 \int_{_nI_{ij}^2} \int_{_nI_{ij}^2} \left(\int_0^{\|z_1-z_2\|} 1 \, dt \right)^2 dz_1 \, dz_2$$

$$\leq {}_nB_{ij}^2 \int_{_nI_{ij}^2} \int_{_nI_{ij}^2} \left(2^{\frac{1}{2}}/n \right)^2 dz_1 \, dz_2{}^2 = 2 \, {}_nB_{ij}^2 \, n^{-6} \quad . \quad \square$$

Proof of Proposition VIII.4:

$$\|\nabla f(z)\|_{L^\infty({}_nI_{ij}^2)} = \left\| \frac{d}{dz} \int_{I^2} F(z') g_0(z-z') dz' \right\|_{L^\infty({}_nI_{ij}^2)}$$

$$= \left\| \int_{I^2} F(z') \frac{d}{dz} g_0(z-z') dz' \right\|_{L^\infty({}_nI_{ij}^2)}$$

$$\leq \left| \int_{I^2} F(z')dz' \right| \; \|\nabla g_0\|_{L^\infty(R^2)}$$

$$\leq \left(\int_{I^2} |F(z')|dz' \right) \|\nabla g_0\|_{L^\infty(R^2)}$$

Assuming that F is in $L^1(I^2)$, we can use Hölder's Inequality:

$$\|\nabla f(z)\|_{L^\infty\left(\bigcup_n I_{ij}^2\right)} \leq \|\nabla g_0\|_{L^\infty(R^2)} \|F\|_{L^2(I^2)} \|\chi_{I^2}\|_{L^2(I^2)}$$

The other way is similar. Notice that $f = g * F$ as well. □

Proof of Proposition VIII.5

Let $X = L^2(I^2)$ and $Y = L^2\left(\bigoplus_{j=1}^m \theta_j^1 \right)$. We can establish that $P_{\{\theta\}} : X \to Y$ is continuous with operator norm $M_1 = m^{\frac{1}{2}}$ since

$$P_\theta f(z) = \int f(z + t\theta)dt, \quad \text{for } z \in \theta^1$$

$$|P_\theta f(z)|^2 \leq \int |f(z + t\theta)|dt)^2$$

$$\|P_\theta f\|_{L^2}^2 = \int_{\theta^1} |P_\theta f(z)|^2 \, dz$$

$$\leq \int_{\theta^1} \left(\int |f(z + t\theta)|dt \right)^2 dz$$

$$\leq \left(\int_{\theta^1} \int |f(z + t\theta)| \, \chi_{I^2} \, dt \, dz \right)^2,$$

where χ_{I^2} is the characteristic function on I^2, and Hölder's inequality implies:

$$\leq \left(\|\chi_{I^2}\|_{L^2} \|f\|_{L^2} \right)^2 = \|f\|_{L^2}^2 \quad , \qquad \text{therefore}$$

$$\|P_\theta f\|_{L^2} \leq \|f\|_{L^2} \quad .$$

We have therefore established that $P_\theta : X \to L^2(\theta^\perp)$ has operator norm less than or equal to 1. But $P_{\{\theta\}}$ maps into Y, so:

$$\|P_{\theta_1} \oplus P_{\theta_2} \oplus \ldots \oplus P_{\theta_m}\|_{L^2}^2 = \sum_{j=1}^{m} \|P_{\theta_j}\|^2 \leq m \|f\|_{L^2}^2 \quad ,$$

and $\|P_{\theta_1} \oplus P_{\theta_2} \oplus \ldots \oplus P_{\theta_m}\|_{L^2} \leq m^{\frac{1}{2}} \|f\|_{L^2} \quad . \qquad \square$

CHAPTER IX

A SPECIAL CASE: IMPROVING THE EMI HEAD SCANNER

This chapter presents a subtle change in the collection of projection data which can provide improved resolution in a reconstruction without requiring improved resolution of the projection data. We present this refinement in the context of the state of the art through the example of the first scanner, that built by the EMI company. The need for uniquely determined reconstructions as explained in earlier chapters is also stressed here. To the knowledge of the author, the theory of the accuracy of the output of existing x-ray scanners has yet to be placed on a sound mathematical basis.

Various parameters which describe the data collection and the reconstruction in the case of the EMI head scanner will now be considered and compared with the parameters given by Theorem 2. The EMI machine reads its projection data at a 1 mm resolution, and its reconstructions have pixels which are 1 mm on a side [24]. Theorem 2 indicates that it is possible to produce reconstructions with a resolution of about 4 mm in these same circumstances since:

   a. the available resolution in projection data is about 1 mm,
   b. the diameter of a human head is about 15 cm,
   c. 4 mm x 4 mm pixels can be used to represent a human head on a 37 x 37 grid, and
   d. a 37 x 37 grid requires projections at angles which in turn require about four times better resolution in the projection data than will be displayed in the reconstruction. (See Chapter VI, table on page 78.) That is, the ratio of the resolution in the reconstruction to the resolution required in the data is about four to one in this case.

We will show that the four to one ratio dictated by Theorem 2 can be improved to such an extent as to get reconstructions which have the same resolution as those produced by the EMI Head Scanner. Moreover, we will demonstrate that the accuracy of such reconstructions can be estimated.

Figure IX.1

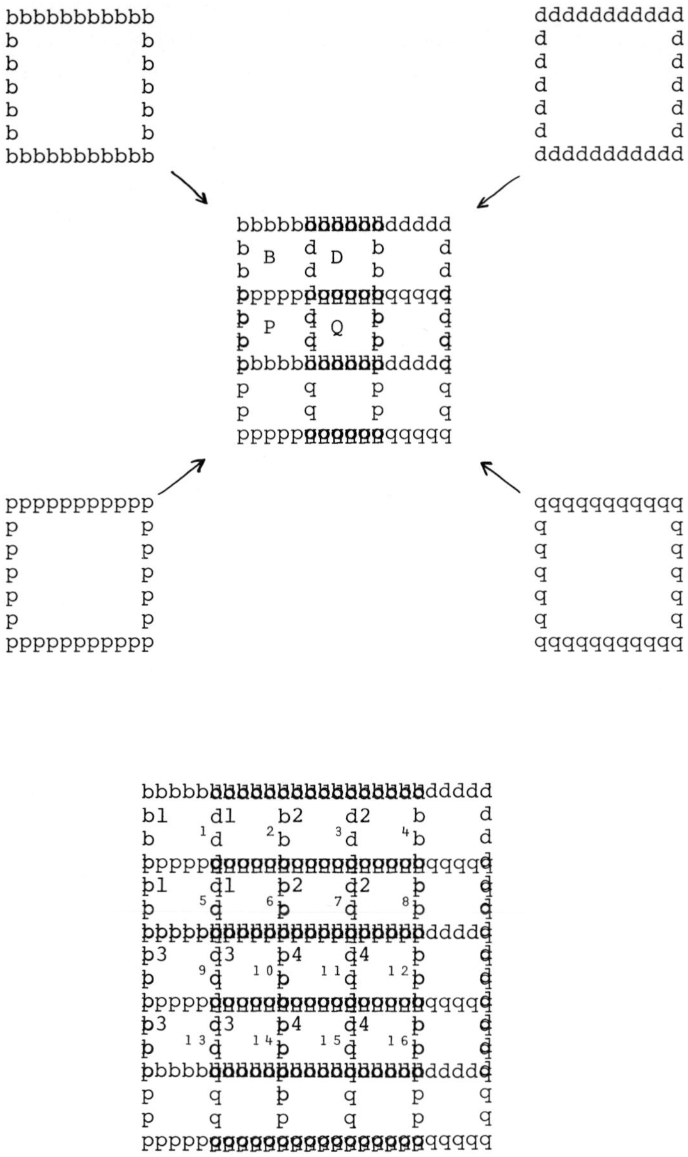

The Use of Purposefully Displaced Reconstructions

In this and the next two sections we will see that it is possible to improve the (picture) resolution of a uniquely determined reconstruction by a factor of two without requiring finer resolution in the data. The technique is to create multiple purposefully displaced reconstructions of a particular object from corresponding multiple sets of projection data. Consider the first diagram in Figure IX.1. Here, four reconstructions, each on a 1 x 1 grid, are overlayed in a purposefully displaced manner. By labeling with letters which are naturally "flagged" - -

- b   corresponds to the upper left reconstruction,
- d   corresponds to the upper right reconstruction,
- p   corresponds to the lower left reconstruction, and
- q   corresponds to the lower right reconstruction,

the particular reconstructions can be distinguished.

The information available in the four relatively coarse reconstructions can be used to obtain a new, 2 x 2 reconstruction with resolution improved by a factor of two. This can be done by assuming that the reconstruction labeled with b's represents the entire unknown objective function -- then, of the nine subsquares created by the boundaries of reconstructions b, d, p, and q, it can be assumed that only the four squares labeled B, D, P, and Q are to be non-zero. This provides us with a system of four equations and four unknowns:

$$B + D + P + Q = 4b$$
$$D + Q = 4d$$
$$P + Q = 4p$$
$$Q = 4q \quad \text{which is clearly non-singular.}$$

The same reasoning can be employed when the second diagram is considered. In this case b is a 2 x 2 grid, we have 4 times 4 or 16 knowns and the same number of unknowns, and it is clear from Figure IX.2 that this is also a non-singular system.

Theorem 2 Applied to Four Sets of Purposefully Displaced Projection Data

The following example demonstrates the method. Theorem 2 [p. 78] indicates that, h_{37}, a reconstruction on a 37 x 37 grid, can be uniquely determined when the projection angles $\{\theta | \tan\theta = \frac{p}{q}$, for $p, q \in Z$, and $|p|, |q| \leq 3\} \cup \{\arctan\frac{1}{4}, \arctan 4, \arctan-\frac{1}{4}, \arctan-4\}$ are used. If we

Figure IX.2

$$\begin{pmatrix} \frac{1}{4} & \frac{1}{4} & 0 & 0 & \frac{1}{4} & \frac{1}{4} & 0 & 0 & 0 & 0 & 0 & 0 & 0 & 0 & 0 & 0 \\ 0 & 0 & \frac{1}{4} & \frac{1}{4} & 0 & 0 & \frac{1}{4} & \frac{1}{4} & 0 & 0 & 0 & 0 & 0 & 0 & 0 & 0 \\ 0 & 0 & 0 & 0 & 0 & 0 & 0 & 0 & \frac{1}{4} & \frac{1}{4} & 0 & 0 & \frac{1}{4} & \frac{1}{4} & 0 & 0 \\ 0 & 0 & 0 & 0 & 0 & 0 & 0 & 0 & 0 & \frac{1}{4} & \frac{1}{4} & 0 & 0 & \frac{1}{4} & \frac{1}{4} \\ 0 & \frac{1}{4} & \frac{1}{4} & 0 & 0 & \frac{1}{4} & \frac{1}{4} & 0 & 0 & 0 & 0 & 0 & 0 & 0 & 0 & 0 \\ 0 & 0 & 0 & \frac{1}{4} & 0 & 0 & 0 & \frac{1}{4} & 0 & 0 & 0 & 0 & 0 & 0 & 0 & 0 \\ 0 & 0 & 0 & 0 & 0 & 0 & 0 & 0 & 0 & \frac{1}{4} & \frac{1}{4} & 0 & 0 & \frac{1}{4} & \frac{1}{4} & 0 \\ 0 & 0 & 0 & 0 & 0 & 0 & 0 & 0 & 0 & 0 & \frac{1}{4} & 0 & 0 & 0 & 0 & \frac{1}{4} \\ 0 & 0 & 0 & 0 & \frac{1}{4} & \frac{1}{4} & 0 & 0 & \frac{1}{4} & \frac{1}{4} & 0 & 0 & 0 & 0 & 0 & 0 \\ 0 & 0 & 0 & 0 & 0 & 0 & \frac{1}{4} & \frac{1}{4} & 0 & 0 & \frac{1}{4} & \frac{1}{4} & 0 & 0 & 0 & 0 \\ 0 & 0 & 0 & 0 & 0 & 0 & 0 & 0 & 0 & 0 & 0 & 0 & \frac{1}{4} & \frac{1}{4} & 0 & 0 \\ 0 & 0 & 0 & 0 & 0 & 0 & 0 & 0 & 0 & 0 & 0 & 0 & 0 & 0 & \frac{1}{4} & \frac{1}{4} \\ 0 & 0 & 0 & 0 & 0 & \frac{1}{4} & \frac{1}{4} & 0 & 0 & \frac{1}{4} & \frac{1}{4} & 0 & 0 & 0 & 0 & 0 \\ 0 & 0 & 0 & 0 & 0 & 0 & 0 & \frac{1}{4} & 0 & 0 & 0 & \frac{1}{4} & 0 & 0 & 0 & 0 \\ 0 & 0 & 0 & 0 & 0 & 0 & 0 & 0 & 0 & 0 & 0 & 0 & \frac{1}{4} & \frac{1}{4} & 0 \\ 0 & 0 & 0 & 0 & 0 & 0 & 0 & 0 & 0 & 0 & 0 & 0 & 0 & 0 & 0 & \frac{1}{4} \end{pmatrix} \begin{pmatrix} r_1 \\ r_2 \\ r_3 \\ r_4 \\ r_5 \\ r_6 \\ r_7 \\ r_8 \\ r_9 \\ r_{10} \\ r_{11} \\ r_{12} \\ r_{13} \\ r_{14} \\ r_{15} \\ r_{16} \end{pmatrix} = \begin{pmatrix} b1 \\ b2 \\ b3 \\ b4 \\ d1 \\ d2 \\ d3 \\ d4 \\ p1 \\ p2 \\ p3 \\ p4 \\ q1 \\ q2 \\ q3 \\ q4 \end{pmatrix}$$

$$\begin{pmatrix} 4 & 4 & 4 & 4 & -4 & -4 & -4 & -4 & -4 & -4 & -4 & -4 & 4 & 4 & 4 & 4 \\ 0 & -4 & 0 & -4 & 4 & 4 & 4 & 4 & 0 & 4 & 0 & 4 & -4 & -4 & -4 & -4 \\ 0 & 4 & 0 & 4 & 0 & -4 & 0 & -4 & 0 & -4 & 0 & -4 & 0 & 4 & 0 & 4 \\ 0 & 0 & 0 & 0 & 0 & 4 & 0 & 4 & 0 & 0 & 0 & 0 & 0 & -4 & 0 & -4 \\ 0 & 0 & -4 & -4 & 0 & 0 & 4 & 4 & 4 & 4 & 4 & 4 & -4 & -4 & -4 & -4 \\ 0 & 0 & 0 & 4 & 0 & 0 & -4 & -4 & 0 & -4 & 0 & -4 & 4 & 4 & 4 & 4 \\ 0 & 0 & 0 & -4 & 0 & 0 & 0 & 4 & 0 & 4 & 0 & 4 & 0 & -4 & 0 & -4 \\ 0 & 0 & 0 & 0 & 0 & 0 & 0 & -4 & 0 & 0 & 0 & 0 & 0 & 4 & 0 & 4 \\ 0 & 0 & 4 & 4 & 0 & 0 & -4 & -4 & 0 & 0 & -4 & -4 & 0 & 0 & 4 & 4 \\ 0 & 0 & 0 & -4 & 0 & 0 & 4 & 4 & 0 & 0 & 4 & 0 & 0 & -4 & -4 \\ 0 & 0 & 4 & 0 & 0 & 0 & -4 & 0 & 0 & 0 & -4 & 0 & 0 & 0 & 4 \\ 0 & 0 & 0 & 0 & 0 & 0 & 4 & 0 & 0 & 0 & 0 & 0 & 0 & 0 & -4 \\ 0 & 0 & 0 & 0 & 0 & 0 & 0 & 0 & 4 & 4 & 0 & 0 & -4 & -4 \\ 0 & 0 & 0 & 0 & 0 & 0 & 0 & 0 & 0 & -4 & 0 & 0 & 4 & 4 \\ 0 & 0 & 0 & 0 & 0 & 0 & 0 & 0 & 0 & 4 & 0 & 0 & 0 & -4 \\ 0 & 0 & 0 & 0 & 0 & 0 & 0 & 0 & 0 & 0 & 0 & 0 & 0 & 4 \end{pmatrix} \begin{pmatrix} b1 \\ b2 \\ b3 \\ b4 \\ d1 \\ d2 \\ d3 \\ d4 \\ p1 \\ p2 \\ p3 \\ p4 \\ q1 \\ q2 \\ q3 \\ q4 \end{pmatrix} = \begin{pmatrix} r_1 \\ r_2 \\ r_3 \\ r_4 \\ r_5 \\ r_6 \\ r_7 \\ r_8 \\ r_9 \\ r_{10} \\ r_{11} \\ r_{12} \\ r_{13} \\ r_{14} \\ r_{15} \\ r_{16} \end{pmatrix} = h_4$$

$$A \begin{pmatrix} h_2^b \\ h_2^d \\ h_2^p \\ h_2^q \end{pmatrix} = h_4$$

The first matrix equation expresses the relationship between the fine reconstruction $(r_1, r_2, \ldots, r_{16})^t$ and the four coarse reconstructions $(b1,b2,\ldots,d1,d2,\ldots,p1,p2,\ldots,q1,q2,q3,q4)$ which appear as the second diagram in Figure IX.1; this matrix is the inverse of the matrix which appears just below it.

let one side of a pixel have length one, then the resolution in the projection data required need not be smaller than $(1^2 + 4^2)^{-\frac{1}{2}} \approx .2425356$. [p. 56] This means that the ratio of resolution in the projection data to the picture resolution seems to be fixed at around .25 or one to four.

To obtain four 37 x 37 purposefully displaced reconstructions: $h_{37}^b, h_{37}^d, h_{37}^p, h_{37}^q$; one should collect four sets of projection data (totaling 4(20) = 80 projections) in a purposefully displaced manner. (For statistical reasons, this is not the same as interpreting each projection four different times.) The ith set of projection data $(P_{\{\theta\}}f)^i$, can be used to obtain h_{37}^i according to the criteria stated in chapter VIII, i.e., h_{37}^i is chosen so as to minimize

$$|| (P_{\{\theta\}}f)^i - P_{\{\theta\}}h_{37}^i ||_{L^2} \cdot \text{ [p. 117]}$$

(For future reference, let $h_{37}^1, h_{37}^2, h_{37}^3, h_{37}^4$ be a renaming of $h_{37}^b, h_{37}^d, h_{37}^p, h_{37}^q$.)

Estimating the Accuracy of a 74 x 74 Reconstruction, h_{74}

Using the same explanation presented in the second section: h_{74} is uniquely determined by $(h_{37}^1, h_{37}^2, h_{37}^3, h_{37}^4)$. We can express the relationship between $(h_{37}^1, h_{37}^2, h_{37}^3, h_{37}^4)$ and h_{74} in the form of a matrix equation:

$$A \begin{pmatrix} h_{37}^1 \\ h_{37}^2 \\ h_{37}^3 \\ h_{37}^4 \end{pmatrix} = h_{74}.$$

An example of A corresponding to the second diagram of Figure IX.1 is the second matrix in Figure IX.2. In this way, A induces a linear transformation from $R^{4(37)^2}$ into $R^{(74)^2} = R^{5476}$. (It is convenient to notice that the natural Euclidean norm on R^{5476} is the same as $||h_{74}||_{L^2}$ when h_{74} is thought of as an element of $Z(74)$. See chapter III.) Let M be the operator norm of A in the natural Euclidean topology, i.e.,

$$M \left\| \begin{pmatrix} h_{37}^1 \\ h_{37}^2 \\ h_{37}^3 \\ h_{37}^4 \end{pmatrix} \right\|_{L^2} \geq ||h_{74}||_{L^2} .$$

If we let $\bar{f}_n$, M_1, and M_2 be as defined in chapter VIII and put $k_n = \bar{f}_n$, then

$$|| f - h_{74} || \leq || f - k_{74} || + || k_{74} - h_{74} ||$$

$$\leq || f - k_{74} || + M \left|\left| \begin{pmatrix} k_{37}^1 \\ k_{37}^2 \\ k_{37}^3 \\ k_{37}^4 \end{pmatrix} - \begin{pmatrix} h_{37}^1 \\ h_{37}^2 \\ h_{37}^3 \\ h_{37}^4 \end{pmatrix} \right|\right| \quad *$$

$$= || f - k_{74} || + M \left(\sum_{i=1}^{4} || k_{37}^i - h_{37}^i ||^2 \right)^{\frac{1}{2}}$$

$$\leq || f - k_{74} || + M M_2 \left(\sum_{i=1}^{4} || P_{\{\theta\}} k_{37}^i - P_{\{\theta\}} h_{37}^i ||^2 \right)^{\frac{1}{2}}$$

$$\leq ||f - k_{74}|| + MM_2 \left(\sum_{i=1}^{4} (|| P_{\{\theta\}} k_{37}^i - (P_{\{\theta\}}f)^i || + || (P_{\{\theta\}}f)^i - P_{\{\theta\}} h_{37}^i ||)^2 \right)^{\frac{1}{2}}$$

$$< ||f - k_{74}|| + MM_2 \left(\sum_{i=1}^{4} (M_1 || k_{37}^i - f || + || (P_{\{\theta\}}f)^i - P_{\{\theta\}} h_{37}^i ||)^2 \right)^{\frac{1}{2}}.$$

We have shown that h_{74}
  a) can be uniquely determined,
  b) requires projection data at a resolution of 1 mm, and
  c) can be tested empirically for its accuracy.

Obtaining a Uniquely Determined Reconstruction with 1 mm Resolution from 1 mm Resolution Projection Data

The last three sections showed that it is possible to obtain h_{74} whenever four particular h_{37}'s can be obtained. When each h_{37} is chosen to conform to the introductory explanation, the pixels of h_{37} must be 4 mm x 4 mm, and the projection data is read at about 1 mm resolution. Now, using the results of the last three sections, we can get an h_{74} which will have pixels of size 2 mm x 2 mm.

Similarly, we can repeat this process three more times to obtain h_{74}^1, h_{74}^2, h_{74}^3, h_{74}^4 which are displaced in the same directions relative to each other as the b, d, p and q reconstructions were displaced in the first section. In this way we obtain an h_{148}. As before, a calculation can be made to estimate the accuracy of h_{148} to obtain:

$$||f - h_{148}|| \leq ||f - k_{148}|| + M'M_2 \left(\sum_{j=1}^{4} \sum_{i=1}^{4} (M_1 || k_{37}^{ij} - f || + || (P_{\{\theta\}}f)^{ij} - P_{\{\theta\}} h_{37}^{ij} ||)^2 \right)^{\frac{1}{2}},$$

---

*Since A is invertible, $\{h_{37}^1, h_{37}^2, h_{37}^3, h_{37}^4\}$ is determined by h_{74}. The fact that $k_{74} = A(k_{37}^1, k_{37}^2, k_{37}^3, k_{37}^4)^t$ is an easy verification.

where M' is the operator norm of the transformation which takes $[h_{37}^{11}, h_{37}^{12}, \ldots, h_{37}^{21}, \ldots, h_{37}^{44}]^t$ to h_{137} and M_2 is the operator norm relative to determining h_{37} from $P_{\{\theta\}}h_{37}$.

Conclusions

Drawing on the results of chapters VI and VIII and the empirical data in [24], this chapter has shown that the EMI Scanner can be modified to produce reconstructions whose accuracy can be estimated. This redesign requires (20)(4)(4) = 320 projections to be read at a 1 mm resolution. The resulting reconstructions would have a pixel size of 1 mm x 1 mm which affords the same resolution as is currently produced by the EMI head scanner.

It should be noted that this entire discussion has preserved a fixed resolution for projection data and has not delved into statistical or Fourier optics considerations. These considerations would depend on the particular apparatus used as discussed in [9].

CHAPTER X

A GENERAL THEORY OF RECONSTRUCTION FROM PROJECTIONS
AND OTHER MATHEMATICAL CONSIDERATIONS RELATED TO THIS PROBLEM

The last chapter presented a particular estimate which can be used as long as certain choices are made, i.e., $L^2(I^2)$ and etc. However, the theoretical background of that estimate does not depend on the particular choices made in Chapter VIII. A more general formulation is given below:

A General Theory of Reconstruction from Projections

For some set of projection angles $\{\theta\} = \{\theta_j\}_{j=1}^{m}$, let X and Y be Banach spaces* which contain f and its projection data respectively, such that $P_{\{\theta\}}: X \to Y$ defined by $P_{\{\theta\}}f = (P_{\theta_1}f, P_{\theta_2}f, \ldots, P_{\theta_m}f)$ is continuous. Let $\Lambda(n)$ be a particular finite dimensional reconstruction space such that:

(1) $\Lambda(n) \subseteq X$ and
(2) if $h, h' \in \Lambda(n)$ and $P_{\{\theta\}}h = P_{\{\theta\}}h'$ then $h = h'$.

Then if some *a priori* information about f is known which estimates the distance between f and some element of $\Lambda(n)$, we can then get an estimate for $\|h - f\|$ (where $\| \ \|$ is a norm which generates the topology of X) when h is empirically obtained.

---
*A Banach Space is a complete normed vector space.

A pathway to an estimate for $\|h - f\|$ can be described in five steps:

(a) Let M_1 be the operator norm of $P_{\{\theta\}}$.[*]

(b) Because of condition (2) and the fact that $\Lambda(n)$ is finite dimensional, there exists a linear transformation from $P_{\{\theta\}}(\Lambda(n))$ into $\Lambda(n)$ which is an inverse for $P_{\{\theta\}}|_{\Lambda(n)}$. Let M_2 be the operator norm of that transformation.

(c) Let $k_0 \in \Lambda(n)$ such that $\varepsilon_1 \geq \|f - k_0\|$ is known. (This is where *a priori* information is utilized.)

(d) Since h is empirically obtained, $P_{\{\theta\}}h$ and $P_{\{\theta\}}f$ can be compared in the topology of Y, i.e., let $\varepsilon_0 \geq \|P_{\{\theta\}}h - P_{\{\theta\}}f\|_Y$.

(e) Our estimate then results from the previous considerations:

$$\begin{aligned}
\|h - f\| &\leq \|f - k_0\| + \|k_0 - h\| \\
&= \varepsilon_1 + M_2 \|P_{\{\theta\}}k_0 - P_{\{\theta\}}h\|_Y \\
&= \varepsilon_1 + M_2 \left(\|P_{\{\theta\}}k_0 - P_{\{\theta\}}f\|_Y + \|P_{\{\theta\}}f - P_{\{\theta\}}h\|_Y \right) \\
&= \varepsilon_1 + M_2 (M_1 \|k_0 - f\| + \varepsilon_0) \\
&= \varepsilon_1 + M_2 (M_1 \varepsilon_1 + \varepsilon_0).
\end{aligned}$$

---

[*] Let U and V be normed linear spaces and T be a linear transformation from U to V, then the operator norm of T is the minimal M satisfying $\|Tu\| \leq M \|u\|$ for all $u \in U$. (It is easy to prove that M exists if U is finite dimensional or if T is continuous.)

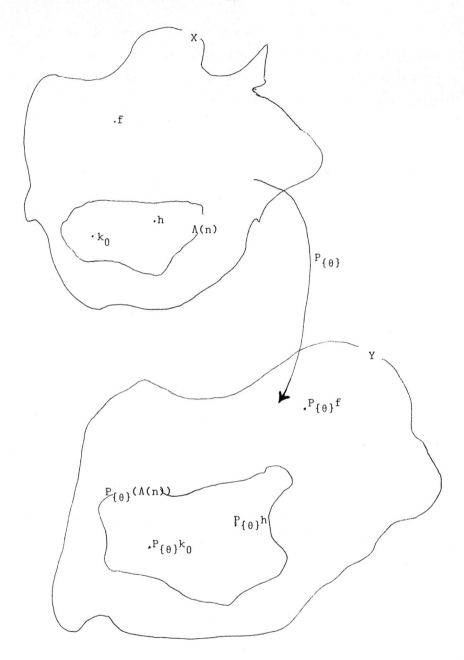

This theory can be regarded as a survey or overview of much of the work done in the field of reconstruction from projections. In particular, two quite different reconstruction techniques can be described by this single theory.

Crowther, De Rosier and Klug [20] solve the reconstruction from projections problem by working in Fourier transform space as explained in Chapter III. They use a different definition of the reconstruction space. Their choices of topologies for X and Y are not clearly delineated in their paper of 1970, but if $X = L_0^2(R^2)$ and $Y = L^2 \left(\bigoplus_{j=1}^{m} \theta_j^1 \right)$, then it is not difficult to recognize that their work is just a special case of this theory. (However, the theory presented in this chapter asks questions that were not adequately addressed in the work of C., De R. and K. For example no estimate of the closeness between h and f is ever given.)

A second reconstruction technique is explained in this thesis and in particular in Chapter VIII. Suffice it to say that most of the questions raised by this theory were answered to a considerable degree of usefulness. The author used the most common choice for the reconstruction space, i.e., $Z(n)$ as defined in Chapter III.

Other Mathematical Considerations Related to Reconstruction from Projections

This section will briefly touch on a series of mathematical questions which are more or less related to the problem of reconstruction from projections. The first few topics are concerned with practical aspects of the problem, while the last question has much less to do with empirical problems.

In the last chapter the infinite dimensional function space which was to contain f, was chosen to be $L^2(I^2)$. However, other L^p spaces could have been used. In fact Proposition VIII.2 can be extended to all of L^p, i.e.,

Proposition X.1 Let $T_n: L^p(I^2) \to L^p(I^2)$ by $T_n f = \bar{f}_n$, where $\bar{f}_n$ satisfies <u>VIII.2</u>, then $T_n \to I$ in the strong operator topology of L^p for $1 \leq p < \infty$.

(The proof of this proposition is actually given as the proof of Proposition VIII.2.)

This remark is made to open the pandora's box of possibilities for the choice of X as defined in the general theory of reconstruction from projections given earlier in this chapter.

Another perplexing question relates to the choice of the "best" projection angles. In Chapter VII, a statistical treatment was given whose goal was to optimize the stability of an estimate of the reconstruction; whereas, Chapter VIII presents a unified theory which can be used to estimate the closeness between h and f. Since this estimate also depends on the choice of the projection angles a controversy can develop. How can these two important goals be simultaneously optimized? That is, how can the statistical and numerical treatments be coordinated so that the reconstruction is optimally stable and is also as close as possible to the unknown function?

This report has said little about the question of the application of uniqueness results to the Fourier methods which are so important to electron microscopy. The fact is that a uniqueness theorem similar to Theorem 2 is needed for an analysis of that problem. Such a result would be quite useful for the experimenters and would be interesting for comparison with the results of this treatise.

An interesting problem which is stimulated by the uniqueness question and K. Smith's theorem relates to the nullspace of $P_{\{\theta_j\}_{j=1}^m}$ for some fixed set of $\{\vec{\theta}_j\}_{j=1}^m \subseteq S^1$. This author has attempted to fully characterize:

$$N \equiv \text{nullspace of } P_{\{\theta\}}\big|_{C_D(R^2)},$$

where D is any compact subset of Euclidean two space. (N has been the subject of other analysis; in fact, $N^\perp$, which is the set of all functions which are perpendicular to everything in N as defined in the L^2 topology, has been characterized by D. Solomon. [73]) In this case, however, our results are limited to finding a fairly large subset of N and a somewhat unsatisfying description of N. It is easy to see that P_{θ_i} is a continuous transformation from $C_D(R^2)$ to $C(\theta_i^1)$, therefore, *N is uniformly closed.* *

Let M be the uniform closure of the space generated by elements of the form given for K defined according to Proposition II.1. It is clear that $M \subseteq N$. Since

$$N = \bigcap_{j=1}^{m} \text{nullspace of } P_{\theta_j}|_{C_D(I^2)},$$

we can now consider the question of characterizing the individual nullspace of each P_{θ_i}. (If we let $t_{\vec{\theta}} f(z) \equiv f(z + t\vec{\theta})$, then:)

Proposition X.2 The nullspace of $P_\theta: C_D(R^2) \to C(\theta^1)$ is the uniform closure of $\{f - t_{\vec{\theta}} f: f \text{ and } t_{\vec{\theta}} f \text{ are elements of } C_D(R^2) \text{ and } t \in R\}$.

Combining these results we get that

i) $N = $ the nullspace of $P_{\{\theta\}}$ is uniformly closed.

ii) $N = \bigcap_{i=1}^{m} \begin{smallmatrix}\text{uniform}\\\text{closure}\\\text{of}\end{smallmatrix} \{f - t_{\vec{\theta}} f: f, t_{\vec{\theta}_j} f \in C_D(R^2), \text{ and } t \in R\}$

iii) $N \supseteq M$ $\begin{smallmatrix}\text{uniform}\\\text{closure}\\\text{of}\end{smallmatrix}$ {subspace of $C_D(R^2)$ generated by elements of the form, K}

---
*"Uniformly closed" means closed in the topology of uniform convergence, i.e., the topology specified by $\|f\| = \sup_{\text{domain } f} |f(z)|$.

This author conjectures that N = M . It certainly seems likely that we are on the verge of completely characterizing N .

(The rest of the chapter presents the proof of Proposition X.2 which requires one lemma.)

Lemma   Let $\lambda: C_D(R) \to R$ by $\lambda(f) = \int f(t)dt$, for D compact. Then the nullspace of λ =

$$\text{uniform closure of } \{f - {}_tf : f \text{ and } f - {}_tf \in C_D(R) , t \in R\}$$

(Here ${}_tf(s) = f(t+s)$.)

Proof: Let $f \in$ kernel of λ, and let $G(t) = \int_{-\infty}^{t} f(s)ds$, then by the fundamental theorem of calculus we have that,

$$\lim_{h \to 0} \frac{G(t + h) - G(t)}{h} = f(t) ,$$

and that this convergence is uniform. Although $G \in C_D(R)$, ${}_hG(t) = G(t + h) \notin C_D(R)$ in general. When f is in the nullspace of λ a further trick can be used: Let $f^+ = f$ when f is positive and zero otherwise, and let $f^- = f - f^+$. Since f is in the nullspace of λ we have:

$$\int f^+ = - \int f^- .$$

Let $n \in Z^+$ then there exists an $r \in R^+$ such that if $A(n) = \{x \in D: f^+(x) \geq 1/n\}$ and $B(r) = \{x \in D: f^-(x) \leq -1/r\}$ then

$$\int_{A(n)} f^+ = - \int_{B(r)} f^- .$$

Now define a sequence of functions f_n which converges uniformly to f:

$$f_n = \begin{cases} f^+ - 1/n & \text{when } f \geq 1/n \\ f^- + 1/r & \text{when } f \leq -1/r \\ 0 & \text{otherwise,} \end{cases}$$

then $f_n \in C_D(R)$ for all n. Define $G_n(t) = \int_{-\infty}^t f_n(s)ds$. It is easy to see that there exists a subsequence of

$$\left\{ m(G_n - {}_mG_n) \right\}_{n=1, \ m=1}^{\infty, \ \infty}$$

contained in the nullspace of λ which converges uniformly to f. $\square$

Proof of Proposition X.2: If we adjust the proof of this Lemma slightly we can find the nullspace of P_{θ_j} for each j. The trick is to notice that although the sequence $\{f_n\}$ is contained in the nullspace of λ, this was not necessary to the proof. We could have let $r = n$ each time. Then the definition of G_n becomes:

$$G_n(z, t) = \int_{-\infty}^t f_n(z + s\vec{\theta})ds \text{ for } z \in \theta^\perp.$$

$\square$

APPENDIX - MEDICAL CONTEXT OF RECONSTRUCTION FROM PROJECTIONS

Tomography is a radiographic diagnostic technique which attempts to improve visualization of certain regions by increasing our ability to localize areas of the body. Conventional or geometric tomography commonly refers to the original techniques by which x-ray source and photographic plate are moved in opposite directions within parallel planes in a coordinated manner. For example, if the source and the film are moved at the same speed, then a planar section of the patient midway between them will have a projection which moves with the film. This planar section will remain in focus on the tomogram while the rest of the intervening tissues are blurred. Conventionally, the coordinated motion of source and film is along a line. Predictably, the region of the body which stays in focus and has a particular *section thickness,* diminishes as the magnitude of the angular displacement *(tomographic angle)* increases. Although a tomogram usually provides quite good spatial resolution, the image contrast diminishes with increasing patient thickness or increasing tomographic angle owing to blurring. Spurious or *phantom images* caused by the blurring arise. Such imaging and localization problems have encouraged the use of other planar paths, including circular, elliptical, spiral, or hypocycloidal. In conventional tomography contrast has remained poor, but its very good spatial resolution maintains the usefulness of conventional tomography in high contrast (bone, airways or contrast media) studies. [53,77]

To improve image contrast, it has been necessary to inject air or iodide substances, creating artificial differences in x-ray attenuation in order to image the ventricles, subarachnoid spaces and blood vessels more accurately. Other methods to improve imaging were explored by Dr. D. E. Kuhl who used a computer to process data collected by a gamma detector at various axial angles after injected radionuclides localized a tumor. [48] His output image, actually a reconstruction, was of a cross-sectional planar section which contained the paths of the detected gamma rays. Since no overlying structures were sampled by radiation, they could not blur the image; nevertheless, Kuhl obtained only passable resolution. An earlier proponent of the use of computer-processed axially collected data, Dr. W. H. Oldendorf, added a collimated scintillation counter. He recognized the potential to the study of biological systems - especially the brain. [58] However, it was not until G. N. Hounsfield and EMI Ltd. utilized these ideas with x-rays that *computerized tomography* became firmly established as a medical diagnostic technique. [57,44]

Like geometric tomography, the best use of computerized tomography requires an understanding of the underlying theory -- in this case, *reconstruction from projections*. Although only poorly understood to this day, reconstruction from projections has yielded qualitative guidelines presently in use. In most examples, a transverse planar section is sampled by a beam narrowed to intersect only that section. Projection data collected by a collimated detector or an array of detectors is processed by a computer according to an algorithm. The digital output displayed on a cathode ray tube is an approximation of the distribution of the x-ray attenuation coefficients in the transverse section. [77]

Computerized tomography is basically noninvasive, and provides excellent contrast at modest spatial resolution, but presents a complicated design problem. Currently available devices apply the theory of reconstruction from projections in a way which is open to mathematical criticism.[*] Data collection methods are under the conflicting stresses of image contrast, spatial resolution, scan speed, and dosage requirements; accuracy is further hampered by the nonhomogeneity of the physical phenomena involved. Algorithms are also evolving to handle these changing constraints; speed and stability in the presence of erroneous data are among the critical issues. Even when the output is virtually error free, interpretation of the image may be difficult.

The Interaction of X-rays with Matter

When a monochromatic x-ray beam, of intensity I, passes through an object, the *attenuation*, ΔI, causes fewer x-rays to pass through to the detector. The linear x-ray attenuation coefficient, μ, expresses the instantaneous ratio $\frac{\Delta I/I}{\Delta x}$ at any point along the path of the beam. For a homogeneous object, μ is a constant and the penetrating intensity, I, can be related to the incident intensity I_o by $I = I_o e^{-\mu x}$ where x expresses the thickness of the transmitting object. The result is that intensity of the transmitted beam decreases exponentially as the thickness of the attenuator increases. [84]

Attenuation means reduction in intensity. Only two processes significantly diminish the number of x-ray photons which penetrate the object under the conditions prevelant in computerized tomography: at lower x-ray energies the photoelectric effect is more important; Compton scattering of x-rays by water is nine times as likely as the

---
[*]The problems with the theory of reconstruction from projection are discussed in depth in Chapter II. On the other hand, the algorithms which process the projection data will be discussed later in this appendix.

photoelectric effect if the energy of the incident photon exceeds 57 keV. [66]

True absorption of x-rays is called the photoelectric effect. An impinging photon can be absorbed via the excitation of an electron of an atom -- that is, causing that electron to jump to a higher energy level or even escape from the atom. The photoelectric effect is more probable with inner shell electrons since they can more easily accept the incident energy at x-ray frequencies. The energy required for escape is greater the larger the atomic number (Z). The probability of photoelectric absorption per gram, that is per unit mass, varies as Z^3. For example, since oxygen has Z = 8 and calcium has Z = 20, an electron of calcium is almost sixteen times as likely to photoelectrically absorb an impinging x-ray as is an electron of oxygen. [66,52]

Like the photoelectric effect, Compton scattering can occur as an interaction between a photon and an electron. This process is most probable when a photon interacts with an outer, loosely-bound atomic electron. Similarly, a transfer of energy occurs but the overriding concern is the change in direction of the altered photon. It is the density of the matter, or more precisely, the number of electrons per cubic centimeter, which most accurately predicts the probability of Compton scattering per unit mass. This may partially explain the significantly lower x-ray attenuation of fat to that of protein. [84,52]

Attenuation of the x-ray beam is generally proportional to the number of electrons encountered, since electons are responsible for both effects; but the atomic number of the atoms in the tissue greatly affects the probability of photoelectric absorption. (A scale of attenuation exists; one Hounsfield or EMI unit represents a 0.1 percent change in attenuation relative to that of water.) [52,77,44,57]

Another aspect of the passage of the x-ray beam through living tissue results in a change in the nature of the spectrum of the incident beam, as the x-ray source is not monochromatic. Since photoelectric absorption varies inversely as the cube of the energy of the incident photon, lower energy photons are more likely to be absorbed. This results in a change in the spectrum of the penetrating x-ray beam toward the higher energies. Such high energy photons are characteristically more penetrating or harder. Beam-hardening favors subsequent Compton scattering over photoelectric absorption. The attenuation coefficient expresses the amount of scattering plus absorption per unit length of a particular kind of living tissue. It is generally assumed that the attenuation coefficient observed for any part of the body will be the same for each projection. The reader should suspect that beam-hardening invalidates this assumption mildly. To reduce this error, 4 to 5 mm of aluminum can be used to preharden the beam. [2,84]

Another important aspect of attenuation results from the distribution of scattering angles accompanying the Compton effect. With an x-ray energy of 73 keV and an accelerating potential of 120 kVp (used by the EMI Scanner), more than half of the scattered photons are deflected through an angle of less than 45°. This means that the geometry of data collection can play a major role in the contrast and noise in the data. Some authors report that in conventional radiology over 90% of the photons incident on the film are forwardly scattered (scattered through an angle less than 90°). [66,71]

The Meaning of a Projection

Imagine that an object were composed of various thicknesses of materials each with its characteristic attenuation coefficient: $\mu_1, \mu_2, \mu_3, \ldots, \mu_n$ and particular thickness: $x_1, x_2, x_3, \ldots, x_n$. Then the observed transmitted intensity I of an incident beam with an intensity I_o would be given by:

$$I = I_o e^{-\mu_1 x_1} e^{-\mu_2 x_2} \ldots e^{-\mu_n x_n} = I_o e^{-(\mu_1 x_1 + \mu_2 x_2 + \ldots + \mu_n x_n)}.$$

So that $\mu_1 x_1 + \mu_2 x_2 + \ldots + \mu_n x_n = \ln(I_o/I)$.[*] Computerized tomography treats the x_i's as knowns and calculates the μ_i's. (See Chapter III.) The data in each projection provides the required values of $\ln(I_o/I)$.

Thickness of the Slice

When a three-dimensional object is considered to be a stacked array of two-dimensional planar arrays, rectangular parallelopipeds are usually specified as the uniformily shaped subregions. There are clear advantages in having the thickness of the slice or, equivalently, the height of each of these parallelopipeds, as small as possible. This would provide good spatial resolution in the vertical direction, and help to minimize the partial volume phenomenon (discussed later). However, the available methods for collecting data about a thin region necessarily diminish the number of photons which reach the detectors:

---
[*] This description ignores or averages the energy dependent aspects of attenuation. An energy dependent treatment can be found in [2].

The x-ray beam can be collimated before entering the patient so that only the particular slice is irradiated; even so, a larger, wedge-shaped region is sampled by the beam. If we avoid counting the only slightly scattered x-rays by collimating the beam at the exit face, we further decrease the number of photons reaching the detector. [67,57]

Another difficulty associated with the use of thinner slices is time. If thinner slices are used, more slices may have to be studied. In most cases, thinning the slice only increases the amount of time needed to collect the data for that slice. Therefore, thinning the slice dramatically increases the amount of time necessary for adequate examination of the patient.

Types of Detectors

There are two types of detectors in use: scintillation crystals frequently composed of sodium iodide and multiwire gas chambers frequently filled with xenon. The scintillation detectors produce a signal by allowing x-rays to excite electrons in the crystal. When an electron returns to an empty orbital of the lattice, light is produced. This signal is amplified via an optically coupled photomultiplier which produces the electronic impulse. Since the time required for the crystal to reach an unexcited state can be comparable to the time between adjacent data collections, this "after glow" effect can produce distortions near transitions between bone and air. Multiwire gas chambers act like Geiger counters. When a photon knocks out an orbital electron of xenon, the resulting ions carry an electric current between wires which are kept at a constant potential. This current provides the signal for the data collection. Scintillation counters provide high amplification of the input x-ray intensity; however, they are large in size and relatively costly. The cheaper gas chambers are smaller and can be packed closely together. Currently, scintillation detectors are used in systems which require a small number of detectors, while the gas chambers are incorporated when the design calls for a large number of detectors packed closely together. An important advantage of these photon-electronic detector systems over film is that the spectrum of response available more than matches the variations in the output intensities of the x-ray beam. [77,57,45]

Parallel and Fan-beam Techniques

In a CT system, the gantry is the structure which holds and provides the motion for the x-ray source, the detector system, and various

collimators. The oldest devices employ the parallel technique of data
collection, and require translational and rotational motions, see p. 8.

Following the first generation machines, various types of fan-beam
geometry have been used. The key to the use of fan-beams is a linear
or circular array of detectors. The only collimation applied in such
a system restricts the x-rays to the particular slice of interest.
Within that slice, the x-rays from a small source are allowed to diverge
into a fan-beam. The array of detectors can simultaneously record the
intensities over large portions of the slice -- thereby receiving the
data for many equations at the same time.

Second generation machines employ the fan-beam geometry and utilize
the same translational and rotational motions of both x-ray source and
rigidly connected detector array which were used by the first machines.
New scanners do not employ translational motions at all; they use fan-
beam geometries and gas chamber detectors. Rotation of the gantry col-
lects large amounts of data in very short periods. The realization of
the data for each projection requires the computer to sort out data
collected at many different angular positions: in one case, the x-ray
source and a relatively large number of detectors are rigidly connected;
another data collection method requires an entire circle of detectors.
In the latter case, only the x-ray source moves; the detector circle is
motionless. In their efforts to decrease scanning time, designers of
these devices have had to make sacrifices regarding calibration, cost,
collimation, inefficient utilization of x-ray dose to patient and, ul-
timately, the accuracy of the reconstruction. [77,68,45]

Resolution of the Data

Since the data must be discrete to be measured experimentally,
some way must be employed to separate the x-ray photons into compartments so that the intensity within each compartment can be measured
independently. In the fan-beam techniques, this may be accomplished
spatially by the use of an array of detectors. The separation of the
compartments can be done chronologically by an internal clock within
the computer or by pulsing the x-ray source as the x-ray gantry is
rotated. It is a different matter in the devices which employ both
translational and rotational motions. In this case a glass plate, on
which is placed a series of parallel, thin strips of lead, is put between the head and the detector. This grid provides the spatial and
chronological separation required for the collation of the photons. [68]

Although each system has a clearly defined distance for the resolution of the collected projection data, the utility of that data depends as well on the expected accuracy. This accuracy can be expressed as *contrast*. Here the reference is made to the precision of the measurement. If σ is the standard deviation of the error in the measurement of the intensity of the emerging x-ray beam when, for example, 150 photons originally entered the patient, then the standard deviation of the measurement when N times 150 photons enter the patient will be $\sigma/\sqrt{N}$. Therefore, if $N = 10^4$ then the standard deviation of the error in the measurement will decrease by a factor of 10^2. This would suggest that if a finer graticule were used, say twice as fine, and everything else remained the same, then the resolution of the data will improve by a factor of two but the contrast of the data will degrade by a factor of $\sqrt{2}$. The situation regarding the trade-off between the contrast and spatial resolution of the fan-beam system is important. Owing to the size of the detectors within the array, resolution is hard to improve; and because of the effects of forward scattering, the standard deviation of the error in the measurements would also be high. On the other hand, time is not critical, since there is no need for translational motion; longer exposure time can allow some improvement in the contrast. [77,53,6]

X-ray Exposure

Maximum exposure to the patient has been measured for the EMI head Scanner and estimated between 2.5 rads per scan to 4 rads for a series of three to four scans; this is comparable to a plain skull series. These exposure rates are consistent with the asymmetric nature of x-ray attenuation -- as explained in the discussion of beam-hardening. For example, 2.8 rads has been quoted as the maximum dosage to the right eye (the one closer to the x-ray tube) in an orbital scan. The left eye may only have one fifth the exposure. [53]

Miscellaneous Aspects of Data Collection

In the first and second generation machines (incorporating rotational and translational motions) the number of projection angles as well as the specification of the particular angles themselves has been a moderate design problem. Lead by some theoretical studies and experimentation, Hounsfield and EMI decided upon the use of equally spaced projection angles and chose 180 for the number of these angles. Recently, EMI has decided to utilize one third this number of angles for certain optional modes of operation. This would reduce scanning time, but only further experimentation can recommend the quality of the produced reconstruction under these conditions. [45]

Scanning time is often considered very important. The advantages of brief scanning include improved versatility, if patient motion is a problem, and more efficient use of the scanner. The EMI head Scanner yields two images (through the use of two detectors collecting photons in adjacent vertical slices) in a total scanning time of about 270 seconds. The use of fan-beam geometry allows much improved scanning time. The second generation scanners have succeeded in cutting scan time to between 150 and 20 seconds and recently marketed scanners have scanning times under ten seconds, but with faster scanning comes loss of accuracy in calculating attenuation coefficients. [77]

Calibration of the source - detector system is of fundamental importance. Since the observed attenuation must allow for the distinction between various tissues, even the slightest instability in the output of the x-ray source or the sensitivity of the detector system can lead to important misinterpretations. The EMI Scanner literally recalibrates itself during each projection through the use of a plastic rod. By assuming that the NaI crystal and the coupled photomultiplier as well as the x-ray source are operating consistently during the data acquisition for a single projection, and given that the x-ray attenuation of the plastic rod is known, the computer can rescale the projection data for that projection angle so that it is consistent with the known information. Similar rescaling procedures are available for the second generation, fan-beam machines. But frequent rescaling of the source-detector system is difficult and/or time consuming for the third generation machines, and this may result in the production of inaccurate reconstructions. [45,24]

A problem receiving recent attention involves the hardware device which converts the analog electronic signal generated by the detector to a form which is (digital) computer readable. The problem is one of accuracy of conversion. The resulting error can be around 0.5% -- a value which exceeds statistically estimated requirements of about 0.02% error in the projection data. However, this source of error can be reduced to negligible levels with proper implementation. [12,7]

The EMI Example

The EMI Scanner employs another mechanism to facilitate accurate data collection. The patient's head is inserted into a cavity fitted with a water bag so that air near the head can be forced away by water pressure. This water arrangement serves at least three purposes. 1) The water tends to preharden the beam, 2) Since water instead of air is adjacent to bone, the detectors are not required to measure very great differences in attenuation over short time intervals during the translational sweep required for each projection -- a problem that tests the accuracy of response of the detection instruments. 3) The outer container for the water allows the beam to traverse a fixed length (27 cm) of water-like attenuating matter -- this further diminishes the stress on accurate response of the detector system.

The gantry of the EMI Scanner uses a conventional x-ray tube energized to about 120-140 kVp with a current of 30 mA. The beam produced is collimated to a rectangle 26 mm tall and a few millimeters wide. At the receiving end is a pair of NaI scintillation detectors coupled to photomultipliers, each responding to half of the beam. A graticule placed right before the detectors collates the photons so that 240 transmission readings are stored by each detector during each translational sweep. The entire gantry rotates 1° and repeats the process in the same plane. This is repeated for 180° taking about 4.5 minutes and produces two CT scans: 13 mm thick adjacent cuts. [57,44]

Algorithms

Although fundamentally different algorithmic processes have been explored, the basic idea behind the prominent methods for passing from radiological projection data to a reconstruction is described in the *back-projection* method. Also called the *summation* method, this technique produces a reconstruction by overlaying a series of spread out or back-projected images which can be obtained from the data in each projection; that is, the value at each point in the reconstruction is the sum of the densities of each ray passing through that point. For example, if a test object consists of a single disk in an homogeneous background and four projections of it were made, the reconstruction would be similar to an eight-point star [see Figure A-1]. Clearly, the back-projection method is not accurate since the original disk has been spread out in eight directions -- indicative of the "point spread" nature of this method. In the example of the reconstruction of the disk, the simple nature of the original object is reflected in the single star pattern created, but when more general objects are imaged with this technique, spurious details may be hard to recognize and discount. Another problem with this method is that the reconstruction does not have projection data which agree with the original data provided. [16,10,23,76]

To improve the back-projected image, spurious details can be avoided by a process which removes the star-shaped or point-spread effects; such an algorithm is named *convolution* or *filtered back-projection*. Alternatively, it is possible to concentrate on the production of a reconstruction whose projection data agree with the original data; algebraic reconstruction technique *(ART)* is the best known algorithm based on this idea. Both kinds of algorithms utilize the back-projection process, but only after they employ rules to alter the projection data. The key difference lies in the way the data are transformed. ART proceeds iteratively by altering the data for each projection angle in a way which depends on the present guess for the reconstructed image, so that when this data is then back-projected, the new image agrees with the original data at that projection angle. [Figure A-2]. Convolution, however, uses a simple rule to change or filter all of the projection data uniformly before any back-projection proceeds. [Figure A-3]. [23]

Originally, the EMI Scanner programmed its computer to calculate a reconstruction using the ART algorithm perhaps because of the following: 1) when the data is error free, ART tends to pick the reconstruction (from the reconstruction space of 160 x 160 arrays of real numbers) which agrees with the projection data and is closest (in the least sum of squared deviation sense) to the zero array, i.e., the 160 x 160 array of zeros. That is, it is not in the nature of ART to introduce spurious images when good data are available. Nothing so clear cut has been deduced about the convolution algorithm; in fact, the reconstruction produced by convolution may not even tend to agree with the provided projection data. 2) Compared to convolution, ART is less affected by noisy data in its choice of reconstruction as long as the noise level exceeds a modest figure. 3) ART's mathematical description is flexible enough to accept the introduction of addition information (called constraints); this flexibility has allowed the creation of various other algorithms in the same family: multiplicative ART, ART2, ART3, ARTIST, SIRT (simultaneous iterative reconstruction technique) and LSIT (least squares iterative reconstruction technique). 4) The logic which justifies ART is robust, for example it allows the use of projection data which does not span a full semicircular arc; this is definitely not the case for convolution. 5) Generally, ART has the capability to more accurately depict large local fluctuations in attenuations without a concomitant amplification of the effects of noise in the data than its rival, convolution. [41,40,34]

With the demand for faster scanning and the rise in confidence in data collection, EMI switched to convolution; they may have considered the following reasons. 1) ART is significantly more time consuming than convolution; even authors who advocate ART admit that convolution utilizes only one third of the amount of computer time. This was not an issue when the gantry motions require 4.5 min, but newer machines operate in one third to one twentieth of these times. 2) Because of the iterative nature of ART -- it proceeds by utilizing the projection data at each angle in isolation and then continues to the next angle, continuing for a number of cycles -- and the fact that real projection data will not match any feasible reconstruction (with probability 1.0), the current guess will tend to wander among virtually equivalent reconstructions after a few cycles of iterations. Convolution's two-step nature (first filter all of the data, then back-project) does not present this difficulty. 3) Since ART tends to wander, some criterion must be used to end its execution; this stopping criterion affects the accuracy of the reconstruction but has not been logically linked to that accuracy. Convolution requires no such arbitrary condition.

Other results further suggest the implementation of Convolution. 1) Convolution has been shown under various ideal conditions to be equivalent to another technique called Fourier Filtering; this near equivalence has lent more credence to the algorithm and provided insights into and guidelines for its improvement. 2) Under conditions generally available in current practice, convolution has been shown to be statistically reliable; in particular, with current limits on computer time, convolution performs better than ART in the representation of small fluctuations of brain matter (when such matter is at least a small distance from the skull), when the projection data has low error content. [16,41,10]

New algorithms have been developed. The current emphasis is on speed since fan-beam geometries have greatly reduced scanning times. "Real time" computational speed refers to the amount of time required to complete the computation of a reconstruction on a computer when the speed of data acquisition is recognized as a limiting factor. It is difficult to surpass convolution's time of execution since it is almost "real time constrained". That is, since the eventual effect of the projection data at any angle on the reconstruction is independent of the data collected at other angles, this effect can be calculated while the data are collected at the next projection angles. Some scanners utilize this fact in their display system. New algorithms will probably have to demonstrate improved reliability and/or require less data acquisition in order to replace convolution; in particular, the speed of execution will not be a controlling issue. [43,22]

FIGURE A-1

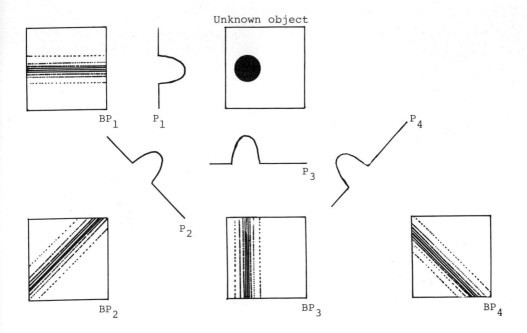

Reconstruction

Illustration of the BACK-PROJECTION reconstruction algorithm. The unknown object is top central. Given information is the set of projection data: P_1, P_2, P_3, P_4. The back-projected images, BP_1, BP_2, BP_3, and BP_4, are the spread out values of the corresponding projections. The reconstruction is obtained by adding together or overlying the back-projected images.

FIGURE A-2

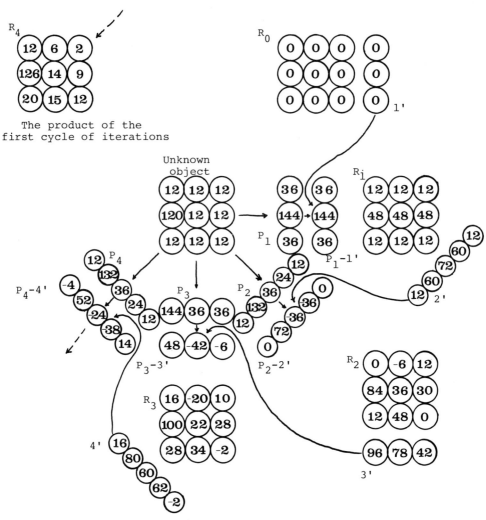

Illustration of the additive ART reconstruction algorithm. The unknown object is central. Given information is the set of projection data: P_1, P_2, P_3, P_4. R_0 is the initial guess at the reconstruction; improved guesses are labeled R_1, R_2, R_3 and R_4. Pseudo projections of the current guess are labeled 1', 2', 3' and 4'. From each projection the values of the corresponding pseudo projection are subtracted; the difference is back-projected among the elements for each ray -- producing the updated guess.

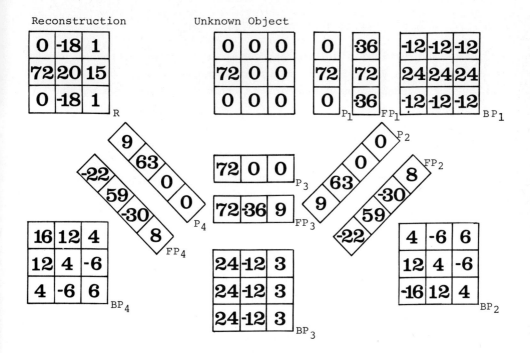

Illustration of the CONVOLUTION reconstruction algorithm. The unknown object is top central. Given information is the set of projection data: P_1, P_2, P_3, and P_4. Filtered projections, FP_1, FP_2, FP_3, and FP_4, are calculated from the corresponding projection data according to a rule:

(1) Set the filtered projection equal to zero. Choose the first non-zero entry of the given projection.
(2) Call that entry "b" and add b to the corresponding entry of the filtered projection.
(3) Add -b/2 to the entrie(s) of the filtered projection one datum removed from where b was added.
(4) Add b/8 to the entrie(s) 2 data removed from where b was added.
(5) Add -b/16 to the entrie(s) 3 data removed from where b was added.
(6) Add b/32 to the entrie(s) 4 data removed from where b was added.
(7) Choose the next non-zero entry of the given projection and go to step (2); if there are no more non-zero entries, stop.

The back-projected images of the filtered projections, BP_1, BP_2, BP_3, BP_4, are superimposed (i.e., added together) to get the reconstruction, R. Convolution performs better when more projection angles are used and when there is less relief.

Representation of a Reconstruction

    A reconstruction is a (finite dimensional) approximation of the x-ray attenuation distribution in a transverse section of the patient. The quality of such an approximation is characterized by two parameters: resolution and accuracy. These terms are applied in the same way as in a description of the information in a photographic record. There is, however, a significant difference between the interpretation of these two kinds of images with respect to the derivation of information at the limits of resolution and accuracy. When a photograph is enlarged so much that the grain size is apparent, the photograph still contains all of the information available in the negative; in particular, if one observes such an image at a greater than normal distance, the high-resolution, spurious details are blurred out and only coarser but informative aspects remain. By making choices which are analogous to choosing the enlarging parameters, a reconstruction is created to depict information at a particular resolution. If the choice of that resolution is inappropriately fine, then, of course, the reconstruction will be unpredictably inaccurate in its representation of details about that size. Rigorous theories place a strict limit on the resolution available from a particular machine design.* If an attempt is made to synthesize an image beyond this resolution, misleading image features may appear. These features will not be removed by viewing the image subsequently at lower resolution. Figure IV.0, p. 69, clarifies this point. This means that careful consideration must go into the choice of the (mathematical) resolution of the reconstruction.

    An important aspect of the presentation of a reconstruction is the use of algorithms to post-process the original reconstruction. These image-enhancing techniques may involve the presentation of the reconstruction on a finer grid than the one used for its original synthesis. These methods remove graininess, smooth curves, take out "inaccuracies", or may even sometimes add "missing" features. The problem with these post-processing schemes is their dubious validity: they imply an improvement in accuracy and resolution. [72]

    The interpretation of a reconstruction is also confused by the non-isotropic resolution presented; for example, a single pixel represents 1.5 x 1.5 x 13 mm of brain matter in an EMI reconstruction. Some authors have described this aspect of computerized tomography as the major technological shortcoming of present devices. So often has this problem caused confusion that it has acquired a special name -

---

*Theorem 2 of chapter VI establishes this fact.

the *partial volume phenomenon*. The typical situation which has caused this type of misinterpretation involves a pixel near the boundary between brain and skull in a cut near the top of the head. If the rectangular parallelopiped intersects both brain and bone tissues, the value for that pixel represents the average attenuation and therefore will be a number intermediate between the normal attenuation values for bone and brain matter. This means that careful interpretation is required in regions where the kind of tissue varies significantly with small changes in depth. Unlike other effects responsible for misinterpretation, the partial volume phenomenon is not responsible for inaccuracies of the image. When this situation is suspected of hiding a dangerous anomaly, contrast can be enhanced by injection of a radioopaque dye followed by collection of another set of projection data. [57]

To provide a measure of accuracy to reconstructions, experiments have been performed using inanimate objects in the place of patients; these objects are usually called *phantoms*. Phantom experiments indicate that reconstructions are statistically likely to accurately represent attenuations within 0.5% of the difference between air and water. Such experiments indicate that accuracy may not be the same everywhere within a particular reconstruction. [44]

To provide a measure of experimentally established resolution, consider various reports from physicians. Anomalies which are 2.0 cm in diameter and exhibit 4 EMI units of contrast are close to the smallest details which can be reliably discriminated. Occasionally, objects as small as 1.0 cm have been observed. However, a careful statistical analysis based on total dosage considerations denies 3 mm resolution and simultaneous 0.5% accuracy; while resolution better than 2 mm can be ruled out owing to the collimator width at the detector. [7,57,9]

The Diagnosis Problem

The problem associated with interpretation of a reconstruction also places constraints on the design of a computerized tomography system. Inaccuracies (called artifacts) of the reconstruction, can result from theoretical, data collection or algorithm short-comings. Misinterpretation of an "accurate" reconstruction by a physician can be expected if his experience or understanding of the interplay between these issues is incomplete. Furthermore, the reliability rating accumulated by doctors provides, at best, only a gross measure of the accuracy of past reconstructions.*

---

*[81] contains a more complete treatment of these issues.

REFERENCES

1. Alta Bates Hospital, Berkeley, use of photograph granted by Ronda Salisbury.

2. Alvarez, R.E. and A. Macovski, "Energy Dependent Reconstruction in X-Ray Computed Tomography," Phys. Med. Biol, 21, (1976) 733.

3. Ambrose, J., "Computerized Transverse Axial Scanning (Tomography): Part 2. Clinical Application," Brit. J. of Rad. XLVI (1973), 1023-1047.

4. _____, "Computerized X-ray Scanning of the Brain," J. Neurosurg. XL (June, 1974).

5. Ayoub, R., Introduction of the Analytic Theory of Numbers, Math Survey #10, AMS (1963).

6. Barrett, H.H., S.K. Gordon, and R.S. Hershel, "Statistical Limitations in Transaxial Tomography," Comput. Biol. Med. 6, (1976) 307.

7. _____ and W. Swindell, "Analog Reconstruction Methods for Transaxial Tomography," Proc. IEEE 65, (1977) 89.

8. Bracewell, R.N. and A.C. Riddle, "Inversion of fanbeam scans in radio astronomy," Astrophys. J. CL, (1967) 427-434.

9. _____, "Correction for Collimator Width (Restoration) in Reconstructive X-Ray Tomography," J. Comp. Asst. Tomo. I(1) (1977).

10. Brooks, R.A. and G. Di Chiro, "Theory of Image Reconstruction in Computed Tomography," Rad. CXVII (Dec., 1975), 561-572.

11. _____ and G. Di Chiro, "Beam hardening in x-ray reconstructive tomography," Physiol. Med. Biol 21, (1976) 390.

12. Budinger, T.F., "Clinical and Research Quantitative Nuclear Medicine System," Med. Radioisotopes Scintigraphy 1972 I, (1973).

13. _____ and G.T. Gullberg, "Three-dimensional reconstruction in nuclear medicine by iterative least-squares and Fourier transform techniques," IEEE Trans. Nucl. Sci. XXI(3) (1974), 2-20.

14. _____, et. al., "Emission Computer Assisted Tomography with Single-Photon and Positron Annihilation Photon Emitters," J. Computer Assisted Tomography I(1) (1977).

15. Cho, Z.H., Ed., SPECIAL ISSUE ON PHYSICAL AND COMPUTATIONAL ASPECTS OF 3-DIMENSIONAL IMAGE RECONSTRUCTION, IEEE Trans. Nucl. Sci. XXI(3) (1974).

16. _____, et. al., "A comparative Study of 3-D Image Reconstruction Algorithms with Reference to Number of Projections and Noise Filtering," IEEE Trans. Nucl. Sc., NS-XXII (Feb., 1975).

17. Colsher, J.G. and R.G. Hart, "Comparison of Algorithms for three dimensional image reconstruction from a series of conical projections," *Image Processing for 2-D and 3-D Reconstruction from Projections Theory and Practice in Medicine and the Physical Sciences*, Stanford, Calif: 1975, pp. Tud2-1-Tud2-4.

18. Cormack, A.M., "Representation of a Function by Its Line Integrals, with Some Radiological Applications," *J. App. Phys.* XXXIV(9) (September, 1963) 2722-2726.

19. _____, "Representation of a Function by its Line Integrals, with Some Radiological Applications. II," *J. App. Phys.* XXXV(10) (October, 1963) 2908-2912.

20. Crowther, R.A., D.J. De Rosier and A. Klug, F.R.S., "The reconstruction of a three-dimensional structure from projections and its application to electron microscopy," *Proc. Roy. Soc. Lond. A.* CCCXVII (1970) 319-340.

21. De Rosier, D.J. and A. Klug, "Reconstruction of three dimensional structures from electron micrographs," *Nature* CCXVII (13 January, 1968) 130-134.

22. Ein-gal, M. "The shadow transform: An approach to cross-sectional imaging," Tech. Rep. 6581-1, Stanford Univ., Stanford, Information Systems Lab, Nov. 1974.

23. Edholm, P. "Image construction in transverse computer tomography," *Acta Radiol* (Supple) 346 (1975) 21.

24. *EMI Scanner Training Manual*, Emitronics, Inc., Middlesex, England, 1975.

25. Fedorov, V.V. *Theory of Optimal Experiments*. New York: Academic Press, 1972.

26. Ferguson, H.R.P., "Reconstruction of Plane Objects by Farey Dissection of Radon's Integral Solution," Post-Deadline Paper, *Image Processing for 2-D and 3-D Reconstruction from Projections: Theory and Practice in Medicine and the Physical Sciences*, Stanford, Calif: 1975, PD6.

27. Frieder, G. and G.T. Herman, "Resolution in reconstructing objects from electron micrographs," *J. Theor. Biol.* XXXIII (1971) 189-211.

28. Gel'fand, I.M., et. al., *Generalized Functions*. Vol. V of 5 vols. New York: Academic Press, 1966.

29. Gilbert, P., "Iterative methods for the reconstruction of three-dimensional objects from projections," *J. Theor. Biol.* XXXVI (1972) 105-117.

30. Glaeser, R., "Limitations to significant information in biological electron-microscopy as a result of radiation-damage," *J. Ultrastr. Res.* XXXVI (1971) 466-482.

31. Goitein, M., "Three-dimensional density reconstruction from a series of two-dimensional projections," *Nucl. Instr. Methods* CI(3) (1971) 509-518.

32. Gordon, R., R. Bender and G.T. Herman, "Algebraic reconstruction techniques (ART) for three-dimensional electron microscopy and x-ray photography," J. Theor. Biol. XXIX (1970) 471-518.

33. _____ and G.T. Herman, "Three-dimensional reconstruction from projections: a review of algorithms," Internat. Rev. of Cytology XXXVIII (1974) 111-151.

34. _____, G.T. Herman and S. Johnson, "Image reconstruction from projections," Scientific American CCXXXIII(4) (October, 1975) 56-61, 64-68. Figures in the text: @ by Scientific American, Inc. all rights reserved.

35. Greenleaf, J.F. and S.A. Johnson, National Bureau of Standards Special Publication 453, Washington, DC (1976) p. 109.

36. Guenther, R.B., et. al., "Reconstruction of objects from radiographs and the location of brain tumors," Proc. Nat. Acad. Sci. USA 71:12 (December, 1974) 4884-4886.

37. Hamaker, C. and D.C. Solomon, "The angles between the null spaces of x-rays," (to appear).

38. Henderson, R. and P.N.T. Unwin, "Three-dimensional model of purple membrane obtained by electron microscopy," Nature CCLVII (4 September, 1975) 28-32.

39. Herman, G.T., A.V. Lakshminarayanan and A. Lent, "Iterative reconstruction algorithms converging to the least-squares solution," Image Processing for 2-D and 3-D Reconstruction from Projections: Theory and Practice in Medicine and the Physical Sciences, Stanford, Calif: 1975, pp. ThA2-1-ThA2-4.

40. _____, A. Lent, and S.W. Rowland, "ART: Mathematics and applications (a report on the mathematical foundations and on the applicability to real data of the algebraic reconstruction techniques)," J. Theor. Biol. XLII (1973) 1-32.

41. _____ and S. Rowland, "Three methods for reconstructing objects from x-rays: a comparative study," Comp. Graph. Image Process. II (1973) 151-178.

42. _____ and A. Lent, "A Family of Iterative Quadratic Optimization Algorithms for Pairs of Inequalities, with Application in Diagnostic Radiology," submitted to Mathematical Programming Studies.

43. _____ and A. Naporstek, "Fast Image Reconstruction Based on a Radom Inversion Formula Appropriate for Rapidly Collected Data," SIAM 101(L) to appear.

44. Hounsfield, G.N. "Computerized Transverse Axial Scanning (Tomography): Part 1. Description of System," Brit. J. of Rad. XLVI (1973) 1016-1022.

45. _____, "Picture Quality of Computed Tomography," Am. J. Roentgenol CXXVII (1976) 3-9.

46. Image Processing for 2-D and 3-D Reconstruction from Projections: Theory and Practice in Medicine and the Physical Sciences; papers given at a conference held by Stanford University Institute for Electronics in Medicine and the Optical Society of America. Richard Gordon, editor. Stanford, Calif: August, 1975.

47. Jones, A.M. and A.M. Fiskin, "Computer Assisted Image Reconstruction of Oligomer Structure," 34th Ann. Proc. Elec. Micros. Soc. Amer., Miami, 1976.

48. Kuhl, D.E. and R.Q. Edwards, "Reorganizing data from the transverse section scans of the brain using digital processing," Radiology 91 (1968) 975.

49. Levinson, R.A. and J.T. Nolte, "Application of algebraic reconstruction techniques to solar coronal x-ray emission," Image Processing for 2-D and 3-D Reconstruction from Projections: Theory and Practice in Medicine and the Physical Sciences, Stanford, Calif: 1975, pp. WA3-1-WA3-4.

50. Logan, B.F. and L.A. Shepp, "Optimal reconstruction of three dimensional structures from electron micrographs," Nature CCXVII (13 January, 1968) 130-134.

51. McCullough, E.C., H.L. Baker and D.W. Houser, "Evaluation of the quantitative and radiation features of a scanning x-ray transverse axial tomograph: the EMI scanner," Radiology 111 (1974) 709.

52. _____, "Photon Attenuation in Computed Tomography," Med. Phys. II:6 (November, 1975).

53. _____ and C.M. Coulam, "Physical and Dosimetric Aspects of Diagnostic Geometrical and Computer-Assisted Tomographic Procedures," Rad. Clinics of N. A. XIV (April, 1976).

54. _____ and J.T. Payne, "X-ray-Transmission Computed Tomography," Med. Phys. IV:2 (March, 1977).

55. Mersereau, R.M. and A.V. Oppenheim, "Digital reconstruction of multidimensional signals from their projections," Proc. IEEE LXII (1974) 1319-1339.

56. Meyer-Ebrecht, D. and W. Wagner, "On the application of ART to conventional X-ray tomography," Image Processing for 2-D and 3-D Reconstruction from Projections: Theory and Practice in Medicine and the Physical Sciences, Stanford, Calif: 1975, pp. TuC3-1-TuC3-4.

57. New, P.F.J., "Computed Tomography: A Major Diagnostic Advance," Hosp. Prac. (Feb. 1975), 55-64.

58. Oldendorf, W.H. "Isolated flying spot detection of radiodensity discontinuities displaying the internal structural pattern of a complex object," IRE Trans. Bio-Med Elect. BME 8 (1961) 68.

59. Oppenheim, B.E., "More accurate algorithm for iterative 3-dimensional reconstruction," IEEE Trans. Nucl. Sci. XXI (1974) 21-43.

60. Proceedings of the International Workshop on Techniques of Three-Dimensional Reconstruction, BNL 20425 (E.B. Marr, ed.) Brookhaven National Laboratory, Upton, NY (1976).

61. Proceedings of the 1975 Workshop on Reconstruction Tomography in Diagnostic Radiology and Nuclear Medicine (San Juan, Puerto Rico), University Park, Baltimore (1977).

62. Radulovic, P.T. and C.M. Vest, "Direct Three Dimensional Reconstruction," *Image Processing for 2-D and 3-D Reconstruction from Projections: Theory and Practice in Medicine and the Physical Sciences*, Stanford, Calif: 1975, pp. TuD1-1-TuD1-4.

63. Ramachandran, G.N. and A.V. Lakshminarayanan, "Three dimensional reconstruction from radiographs and electron micrographs: application of convolutions instead of Fourier transforms," *Proc. Natl. Acad. Sci. USA* LXVIII (1971) 2236-2240.

64. Rao, C. Radhakrishna. *Linear Statistical Inference and Its Applications*. New York: Wiley and Sons, 1965.

65. Rao P.S., E.C. Gregg, "Attenuation of monoenergetic gamma rays in tissues," *Am. J. Roentgenol.* 123 (1975) 631.

66. Ridgeway, A. and W. Thumm, *The Physics of Medical Radiography*. London: Addison-Wesley, 1968.

67. Robinson, A.L., "Image reconstruction (I): computerized X-ray scanners," *Science* CXC (7 November, 1975) 542-543, 593.

68. _____, "Image reconstruction (II): computerized scanner explosion," *Science* CXC (14 November, 1975) 647-648, 710.

69. Shannon, C.E., "Communication in the presence of noise," *IEEE Proc.* XXXVII(1) (1949) 10-21.

70. Shepp, L.A. and B.F. Logan, "The Fourier reconstruction of a head section," *IEEE Trans. Nucl. Sci.* XXI (1974) 21-43.

71. Smith, K.T., D.C. Solomon and S.L. Wagner, "Practical and mathematical aspects of the problem of reconstructing objects from radiographs," *Bull. Amer. Math. Soc.* LXXXIII(6) (November, 1977) 1227-1270.

72. Smith, P.R., et al., "Towards the Assessment of the Limitations on Computerized Axial Tomography," *Neuroradiology* IX (1975) 1-8.

73. Solomon, D.C., "The x-ray transform," *J. Math. Anal. Appl.* (to appear).

74. Sweeney, D.W. and C.M. Vest, *Appl. Opt* 12, (1973) 2649.

75. Swindell, W. and H.H. Barrett, "Computerized Tomography: Taking Sectional X-rays," *Phys. Today* (December, 1977).

76. Takahashi, S., "Rotation Radiography," Japan Soc. Promotion Sci., Tokyo 1957.

77. Ter-Pogossian, M.M. "Computerized cranial tomography: equipment and physics," *Semin Roentgenol* 12 (1977) 13.

78. Tretiak, O.J., "The point-spread function for the convolution algorithm," Image Processing for 2-D and 3-D Reconstruction from Projections: Theory and Practice in Medicine and the Physical Sciences, Stanford, Calif: 1975, p. ThA5-1-ThA5-3.

79. Wald, A., "On the efficiency of statistical investigations," Ann. Math. Stat. XIV (1943) 131-140.

80. Wang, L., "3-D reconstruction algorithms for fanbeam scans," Image Processing for 2-D and 3-D Reconstruction from Projections: Theory and Practice in Medicine and the Physical Sciences, Stanford, Calif: 1975, p. WB6-1-WB6-4.

81. Weisberg, L. and M.B. Katz, Cerebral Computed Tomography - A Text Atlas, Saunders 1978.

82. Wernecke, S.J., "Maximum entropy image reconstruction," Image Processing for 2-D and 3-D Reconstruction from Projections: Theory and Practice in Medicine and the Physical Sciences Stanford, Calif: 1975, p. WA6-1-WA6-4.

83. Yosida, K., Functional Analysis. 3rd ed. New York: Springer-Verlag, 1971, p. 182.

84. Young, M.E.J., Radiological Physics. New York: Academic Press, 1957.

Bio–mathematics

Managing Editors: K. Krickeberg, S. A. Levin

Editorial Board: H. J. Bremermann, J. Cowan,
W. M. Hirsch, S. Karlin, J. Keller, R. C. Lewontin,
R. M. May, J. Neyman, S. I. Rubinow, M. Schreiber,
L. A. Segel

Volume 1:
Mathematical Topics in Population Genetics
Edited by K. Kojima
1970. 55 figures. IX, 400 pages
ISBN 3-540-05054-X

"...It is far and away the most solid product I have ever seen labelled biomathematics."
American Scientist

Volume 2: E. Batschelet
Introduction to Mathematics for Life Scientists
2nd edition. 1975. 227 figures. XV, 643 pages
ISBN 3-540-07293-4

"A sincere attempt to relate basic mathematics to the needs of the student of life sciences."
Mathematics Teacher

M. Iosifescu, P. Tăutu
Stochastic Processes and Applications in Biology and Medicine

Volume 3
Part 1: **Theory**
1973. 331 pages.
ISBN 3-540-06270-X

Volume 4
Part 2: **Models**
1973. 337 pages
ISBN 3-540-06271-8

Distributions Rights for the Socialist Countries:
Romlibri, Bucharest

"... the two-volume set, with its very extensive bibliography, is a survey of recent work as well as a textbook. It is highly recommended by the reviewer."
American Scientist

Volume 5: A. Jacquard
The Genetic Structure of Populations
Translated by B. Charlesworth, D. Charlesworth
1974. 92 figures. XVIII, 569 pages
ISBN 3-540-06329-3

"...should take its place as a major reference work.."
Science

Volume 6: D. Smith, N. Keyfitz
Mathematical Demography
Selected Papers
1977. 31 figures. XI, 515 pages
ISBN 3-540-07899-1

This collection of readings brings together the major historical contributions that form the base of current population mathematics tracing the development of the field from the early explorations of Graunt and Halley in the seventeenth century to Lotka and his successors in the twentieth. The volume includes 55 articles and excerpts with introductory histories and mathematical notes by the editors.

Volume 7: E. R. Lewis
Network Models in Population Biology
1977. 187 figures. XII, 402 pages
ISBN 3-540-08214-X

Directed toward biologists who are looking for an introduction to biologically motivated systems theory, this book provides a simple, heuristic approach to quantitative and theoretical population biology.

Springer-Verlag
Berlin
Heidelberg
New York

A Springer Journal

Journal of Mathematical Biology

Ecology and Population Biology
Epidemiology
Immunology
Neurobiology
Physiology
Artificial Intelligence
Developmental Biology
Chemical Kinetics

Edited by H. J. Bremermann, Berkeley, CA; F. A. Dodge, Yorktown Heights, NY; K. P. Hadeler, Tübingen; S. A. Levin, Ithaca, NY; D. Varjú, Tübingen.

Advisory Board: M. A. Arbib, Amherst, MA; E. Batschelet, Zürich; W. Bühler, Mainz; B. D. Coleman, Pittsburgh, PA; K. Dietz, Tübingen; W. Fleming, Providence, RI; D. Glaser, Berkeley, CA; N. S. Goel, Binghamton, NY; J. N. R. Grainger, Dublin; F. Heinmets, Natick, MA; H. Holzer, Freiburg i. Br.; W. Jäger, Heidelberg; K. Jänich, Regensburg; S. Karlin, Rehovot/Stanford CA; S. Kauffman, Philadelphia, PA; D. G. Kendall, Cambridge; N. Keyfitz, Cambridge, MA; B. Khodorov, Moscow; E. R. Lewis, Berkeley, CA; D. Ludwig, Vancouver; H. Mel, Berkeley, CA; H. Mohr, Freiburg i. Br.; E. W. Montroll, Rochester, NY; A. Oaten, Santa Barbara, CA; G. M. Odell, Troy, NY; G. Oster, Berkeley, CA; A. S. Perelson, Los Alamos, NM; T. Poggio, Tübingen; K. H. Pribram, Stanford, CA; S. I. Rubinow, New York, NY; W. v. Seelen, Mainz; L. A. Segel, Rehovot; W. Seyffert, Tübingen; H. Spekreijse, Amsterdam; R. B. Stein, Edmonton; R. Thom, Bures-sur-Yvette; Jun-ichi Toyoda, Tokyo; J. J. Tyson, Blacksbough, VA; J. Vandermeer, Ann Arbor, MI.

Springer-Verlag
Berlin
Heidelberg
New York

Journal of Mathematical Biology publishes papers in which mathematics leads to a better understanding of biological phenomena, mathematical papers inspired by biological research and papers which yield new experimental data bearing on mathematical models. The scope is broad, both mathematically and biologically and extends to relevant interfaces with medicine, chemistry, physics and sociology. The editors aim to reach an audience of both mathematicians and biologists.

Lecture Notes in Biomathematics

Vol. 25: P. Yodzis, Competition for Space and the Structure of Ecological Communities. VI, 191 pages. 1978.

Vol. 26: M. B. Katz, Questions of Uniqueness and Resolution in Reconstruction from Projections. IX, 175 pages. 1978.

This series reports new developments in biomathematics research and teaching – quickly, informally and at a high level. The type of material considered for publication includes:

1. Preliminary drafts of original papers and monographs
2. Lectures on a new field or presentations of new angles in a classical field
3. Seminar work-outs
4. Reports of meetings, provided they are
    a) of exceptional interest and
    b) devoted to a single topic.

Texts which are out of print but still in demand may also be considered if they fall within these categories.

The timeliness of a manuscript is more important than its form, which may be unfinished or tentative. Thus, in some instances, proofs may be merely outlined and results presented which have been or will later be published elsewhere. If possible, a subject index should be included. Publication of Lecture Notes is intended as a service to the international scientific community, in that a commercial publisher, Springer-Verlag, can offer a wide distribution of documents which would otherwise have a restricted readership. Once published and copyrighted, they can be documented in the scientific literature.

Manuscripts

Manuscripts should be no less than 100 and preferably no more than 500 pages in length.

They are reproduced by a photographic process and therefore must be typed with extreme care. Symbols not on the typewriter should be inserted by hand in indelible black ink. Corrections to the typescript should be made by pasting in the new text or painting out errors with white correction fluid. Authors receive 75 free copies and are free to use the material in other publications. The typescript is reduced slightly in size during reproduction; best results will not be obtained unless the text on any one page is kept within the overall limit of 18 x 26.5 cm (7 x 10½ inches). On request, the publisher will supply special paper with the typing area outlined.

Manuscripts in English, German or French should be sent to Dr. Simon Levin, Section of Ecology and Systematics, 235 Langmuir Laboratory Cornell University, Ithaca, NY 14853/USA or directly to Springer-Verlag Heidelberg.

Springer-Verlag, Heidelberger Platz 3, D-1000 Berlin 33
Springer-Verlag, Neuenheimer Landstraße 28–30, D-6900 Heidelberg 1
Springer-Verlag, 175 Fifth Avenue, New York, NY 10010/USA

ISBN 3-540-09087-8
ISBN 0-387-09087-8